Collins

AQA GCSE (9–1)

Chemistry

Grade 6–7 Booster Workbook

Dorothy Warren
Gemma Young

William Collins' dream of knowledge for all began with the publication of his first book in 1819. A self-educated mill worker, he not only enriched millions of lives, but also founded a flourishing publishing house. Today, staying true to this spirit, Collins books are packed with inspiration, innovation and practical expertise. They place you at the centre of a world of possibility and give you exactly what you need to explore it.

Collins. Freedom to teach.

Published by Collins
An imprint of HarperCollins*Publishers*
The News Building
1 London Bridge Street
London
SE1 9GF

HarperCollins *Publishers*
Macken House 39/40,
Mayor Street Upper,
Dublin 1, D01 C9W8,
Ireland

Browse the complete Collins catalogue at
www.collins.co.uk

10 9 8 7 6 5 4 3 2

ISBN 978-0-00-832255-7

British Library Cataloguing-in-Publication Data
A catalogue record for this publication is available from the British Library.

Authors: Dorothy Warren, Gemma Young
Expert reviewer: Amanda Graham
Development editor: Tim Jackson
Commissioning editor: Rachael Harrison & Jennifer Hall
In-house editor: Alexandra Wells
Copyeditor: Stuart Lloyd
Proof reader: Jan Schubert
Answer checker: Sarah Binns
Cover designers: The Big Mountain Design & Creative Direction
Cover photo: bl: Maksim_Gusev/Shutterstock, tr: Respiro/Shutterstock
Typesetter and illustrator: Jouve India Private Limited
Production controller: Katharine Willard
Printed and bound by: Ashford Colour Press Ltd.

This book is produced from independently certified FSC™ paper to ensure responsible forest management.

For more information visit: www.harpercollins.co.uk/green

The publishers gratefully acknowledge the permission granted to reproduce the copyright material in this book. Every effort has been made to trace copyright holders and to obtain their permission for the use of copyright material. The publishers will gladly receive any information enabling them to rectify any error or omission at the first opportunity.

Contents

Section 1 ● Atomic structure and the periodic table

Section 2 ● Bonding, structure and the properties of matter

Section 3 ● Quantitative chemistry

Section 4 ● Chemical changes

Section 5 ● Energy changes in reactions

Section 6 ● Rates of reaction

Section 7 ● Organic chemistry

Section 8 ● Chemical analysis

Section 9 ● Chemistry of the atmosphere

Section 10 ● Using resources

Answers

Introduction

This workbook will help you build your confidence in answering Chemistry questions for GCSE Chemistry and GCSE Combined Science.

It gives you practice in using key scientific words, writing longer answers, answering synoptic questions as well as applying knowledge and analysing information.

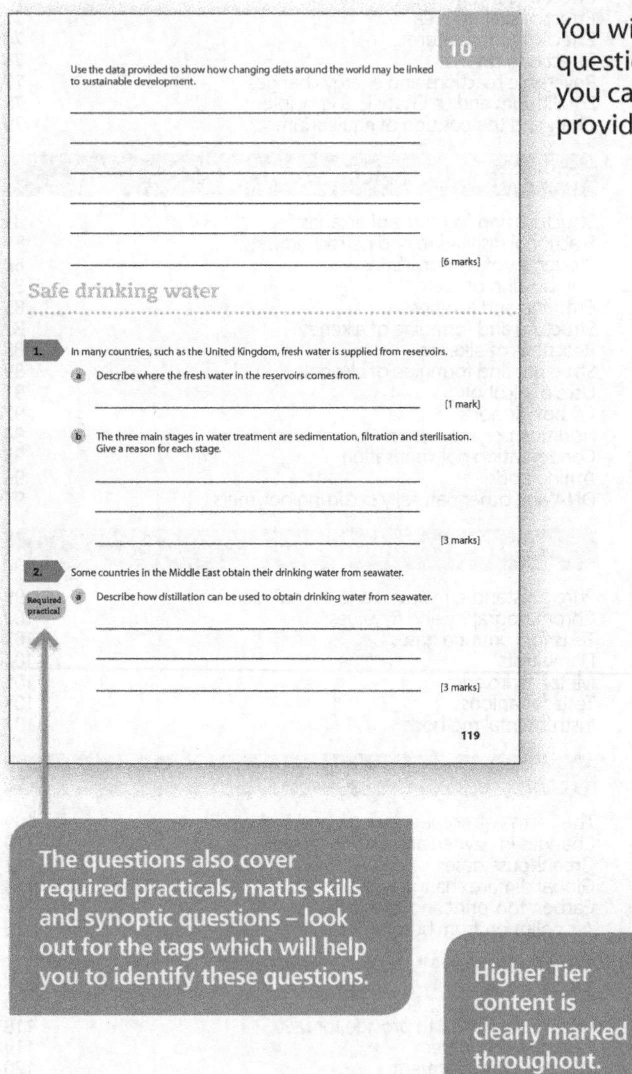

You will find all the different question types in the workbook so you can get plenty of practice in providing short and long answers.

Learn how to answer test questions with annotated worked examples.

This will help you develop the skills you need to answer questions.

The questions also cover required practicals, maths skills and synoptic questions – look out for the tags which will help you to identify these questions.

Higher Tier content is clearly marked throughout.

10

3. Modern lifestyles require organised waste water management systems.

Worked Example: Compare the processing of sewage, agricultural waste water and industrial waste water. [3 marks]

Sewage and agricultural waste water always require the removal of organic waste whereas industrial waste water may not. [1 mark] Sewage and agricultural waste water contain harmful microbes which need to be removed but industrial waste water does not. [1 mark] Instead it contains harmful chemicals that need to be removed. [1 mark]

Suitable connectives for a 'compare' question include 'whereas', 'but', 'instead of' and 'otherwise'.

Alternative methods of extracting metals

1. Useful metals, such as copper, have been mined from ancient times but now alternative methods of extraction are being developed.

Higher Tier only

a Give two reasons why new methods of extraction are being developed.

____

____________________ [2 marks]

b Use words from the box to complete the sentences.

bioleaching	burned	electrolysed	harvested
metal	phytomining	processed	

__________ uses plants to absorb __________ compounds.

The plants are grown, __________ and then __________

to produce ash that contains metal compounds. [4 marks]

2. Copper can be extracted by an alternative method that uses plants.

a Name the copper compound found in the ash after the plant has been burnt. __________ [1 mark]

Worked Example: b Describe how copper metal can be obtained from copper oxide using scrap iron. [3 marks]

Step 1: React copper oxide with dilute sulfuric acid to make copper sulfate solution. [1 mark]

Step 2: Add the iron to the copper sulfate solution. A displacement reaction will take place leaving the copper. [1 mark]

This answer is correct but in step 2 you could use obtain the copper by electrolysis instead of using a displacement reaction.

121

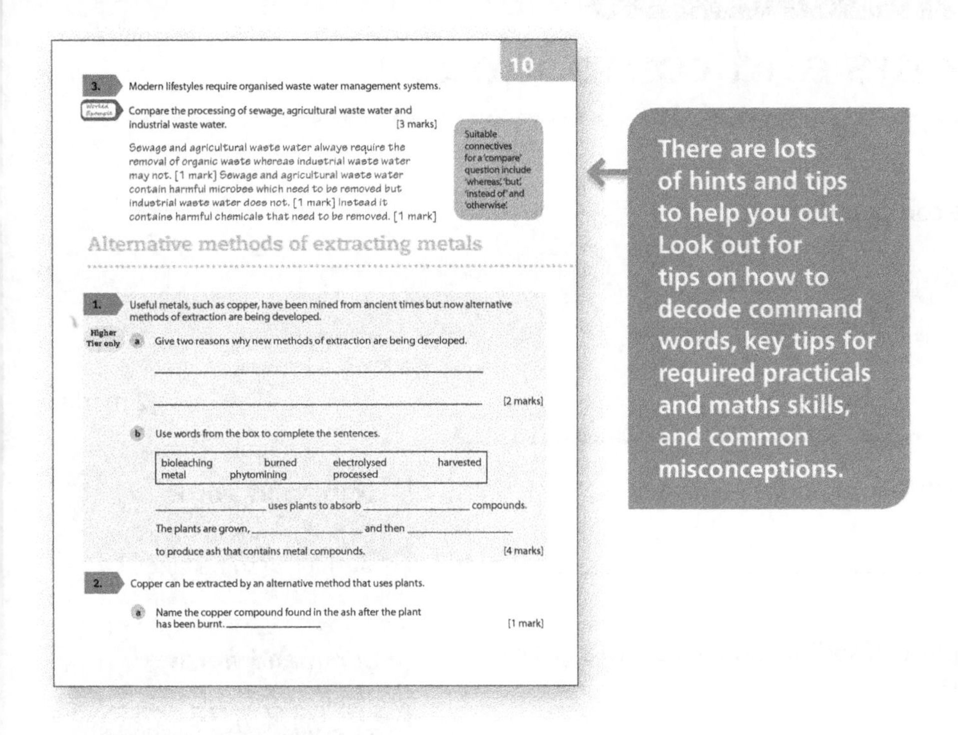

10

3. Modern lifestyles require organised waste water management systems.

Worked Example

Compare the processing of sewage, agricultural waste water and industrial waste water. [3 marks]

Sewage and agricultural waste water always require the removal of organic waste whereas industrial waste water may not. [1 mark] Sewage and agricultural waste water contain harmful microbes which need to be removed but industrial waste water does not. [1 mark] Instead it contains harmful chemicals that need to be removed. [1 mark]

Suitable connectives for a 'compare' question include 'whereas', 'but', 'instead of' and 'otherwise'.

Alternative methods of extracting metals

1. Useful metals, such as copper, have been mined from ancient times but now alternative methods of extraction are being developed.

Higher Tier only

a Give two reasons why new methods of extraction are being developed.

____

____________________ [2 marks]

b Use words from the box to complete the sentences.

bioleaching	burned	electrolysed	harvested
metal	phytomining	processed	

____________ uses plants to absorb ____________ compounds.

The plants are grown, ____________ and then ____________

to produce ash that contains metal compounds. [4 marks]

2. Copper can be extracted by an alternative method that uses plants.

a Name the copper compound found in the ash after the plant has been burnt. ____________ [1 mark]

There are lots of hints and tips to help you out. Look out for tips on how to decode command words, key tips for required practicals and maths skills, and common misconceptions.

The amount of support gradually decreases throughout the workbook. As you build your skills you should be able to complete more of the questions yourself.

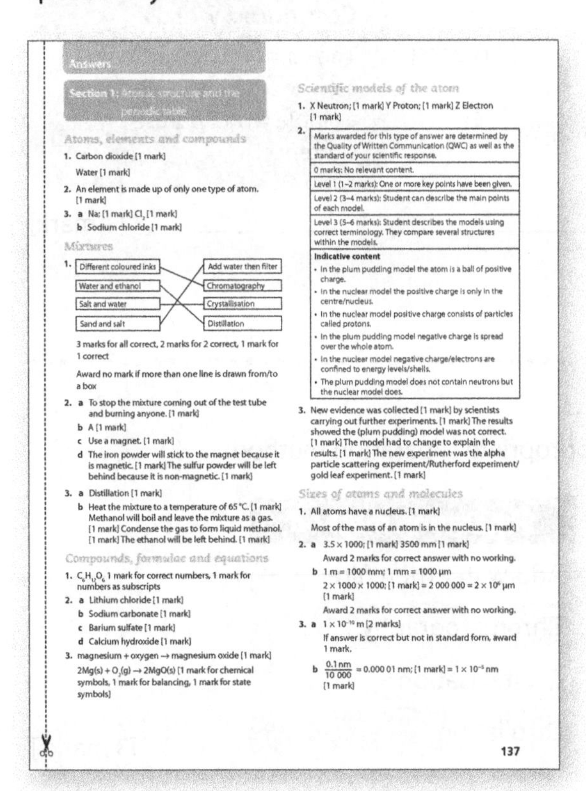

Answers

Section 1: Atomic structure and the periodic table

Atoms, elements and compounds

1. Carbon dioxide [1 mark]
   Water [1 mark]
2. An element is made up of only one type of atom. [1 mark]
3. **a** Na; [1 mark] Cl_2 [1 mark]
   b Sodium chloride [1 mark]

Mixtures

1. Different coloured inks — Chromatography
   Water and ethanol — Distillation
   Salt and water — Crystallisation
   Sand and salt — Add water then filter

   3 marks for all correct, 2 marks for 2 correct, 1 mark for 1 correct

   Award no mark if more than one line is drawn from/to a box
2. **a** To stop the mixture coming out of the test tube and burning anyone. [1 mark]
   b A [1 mark]
   c Use a magnet. [1 mark]
   d The iron powder will stick to the magnet because it is magnetic. [1 mark] The sulfur powder will be left behind because it is non-magnetic. [1 mark]
3. **a** Distillation [1 mark]
   b Heat the mixture to a temperature of 65 °C. [1 mark] Methanol will boil and leave the mixture as a gas. [1 mark] Condense the gas to form liquid methanol. [1 mark] The ethanol will be left behind. [1 mark]

Compounds, formulae and equations

1. $C_6H_{12}O_6$ 1 mark for correct numbers, 1 mark for numbers as subscripts
2. **a** Lithium chloride [1 mark]
   b Sodium carbonate [1 mark]
   c Barium sulfate [1 mark]
   d Calcium hydroxide [1 mark]
3. magnesium + oxygen → magnesium oxide [1 mark]
   $2Mg(s) + O_2(g) \rightarrow 2MgO(s)$ [1 mark for chemical symbols, 1 mark for balancing, 1 mark for state symbols]

Scientific models of the atom

1. X Neutron; [1 mark] Y Proton; [1 mark] Z Electron [1 mark]
2.

Marks awarded for this type of answer are determined by the Quality of Written Communication (QWC) as well as the standard of your scientific response.
0 marks: No relevant content.
Level 1 (1–2 marks): One or more key points have been given.
Level 2 (3–4 marks): Student can describe the main points of each model.
Level 3 (5–6 marks): Student describes the models using correct terminology. They compare several structures within the models.
Indicative content • In the plum pudding model the atom is a ball of positive charge. • In the nuclear model the positive charge is only in the centre/nucleus. • In the nuclear model positive charge consists of particles called protons. • In the plum pudding model negative charge is spread over the whole atom. • In the nuclear model negative charge/electrons are confined to energy levels/shells. • The plum pudding model does not contain neutrons but the nuclear model does.

3. New evidence was collected [1 mark] by scientists carrying out further experiments. [1 mark] The results showed the (plum pudding) model was not correct. [1 mark] The model had to change to explain the results. [1 mark] The new experiment was the alpha particle scattering experiment/Rutherford experiment/gold leaf experiment. [1 mark]

Sizes of atoms and molecules

1. All atoms have a nucleus. [1 mark]
   Most of the mass of an atom is in the nucleus. [1 mark]
2. **a** 3.5×1000; [1 mark] 3500 mm [1 mark]
   Award 2 marks for correct answer with no working.
   b 1 m = 1000 mm; 1 mm = 1000 μm
   $2 \times 1000 \times 1000$; [1 mark] = 2 000 000 = 2×10^6 μm [1 mark]
   Award 2 marks for correct answer with no working.
3. **a** 1×10^{-10} m [2 marks]
   If answer is correct but not in standard form, award 1 mark.
   b $\frac{0.1 \text{ nm}}{10\,000} = 0.000\,01$ nm; [1 mark] = 1×10^{-5} nm [1 mark]

137

There are answers to all the questions at the back of the book. You can check your answers yourself or your teacher might tear them out and give them to you later to mark your work.

Atoms, elements and compounds

1. Which of these are compounds?

Tick **two** boxes.

☐ Air ☐ Carbon dioxide

☐ Iron ☐ Water

[2 marks]

2. Define the word 'element'.

____ [1 mark]

> **Command words**
> 'Define' means to state the meaning of something.

3. A teacher heats a piece of sodium and adds it to a gas jar containing chlorine.

> **Command words**
> When you are asked to 'give' an answer you only need to write a short answer, not an explanation or a description.

a Give the chemical symbols for sodium atoms and chlorine molecules.

Sodium atoms ____________________

Chlorine molecules ____________________ [2 marks]

> **Command words**
> The command word 'name' asks you to answer with a single word, phrase or sentence.

b Name the compound formed in the reaction.

____ [1 mark]

Mixtures

1. Draw **one** line from each mixture to the most appropriate separation method.

Worked Example

Mixture	Separation method
Different coloured inks	Add water then filter
Water and ethanol	Chromatography
Salt and water	Crystallisation
Sand and salt	Distillation

(Worked example: Different coloured inks — Chromatography)

[3 marks]

2. A student carried out the following experiment.

A Measure 1.6 g of iron powder and add it to a test tube.

B Measure 0.4 g of sulfur and add it to the same test tube. Mix well.

C Heat the test tube and contents over a Bunsen burner until an orange glow is seen inside the test tube.

a In step C the student should plug the end of the test tube with mineral wool. Suggest why the student should do this.

______________________________ [1 mark]

Synoptic **b** In which step does the test tube contain a pure element? Tick **one** box.

☐ A ☐ B ☐ C [1 mark]

Command words

When a question asks you to 'suggest' something it expects you to use your understanding in what may be an unfamiliar situation.

c Suggest a method the student could use to separate the mixture formed in step B.

______________________________ [1 mark]

d Explain how the method you gave in part c works.

____

____

______________________________ [2 marks]

Command words

When you are asked to 'explain' something you need to state the reasons for it happening.

3. Ethanol and methanol are both liquids. The boiling point of ethanol is 78 °C. The boiling point of methanol is 65 °C.

a Identify the separating technique you would use to separate a mixture of ethanol and methanol.
Describe how the separating technique works.

____

____

____

____

______________________________ [5 marks]

Compounds, formulae and equations

1. The diagram shows a molecule of fructose, a simple sugar.
Fructose contains carbon, hydrogen and oxygen.

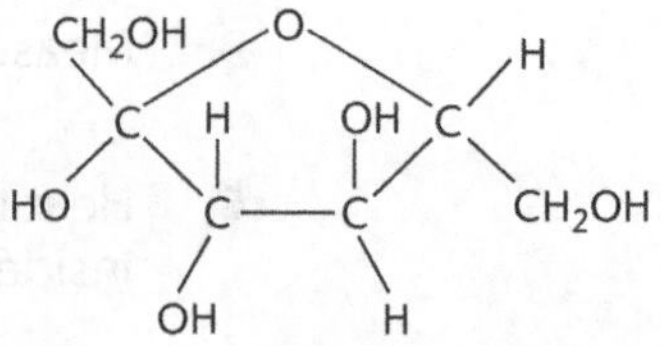

Write the formula of fructose.

____ [2 marks]

2. Name the following compounds.

a LiCl ____ [1 mark]

b Na_2CO_3 ____ [1 mark]

c $BaSO_4$ ____ [1 mark]

d $Ca(OH)_2$ ____ [1 mark]

Here are some tips for writing symbol equations.
Use the periodic table to look up the symbols of elements.
Elements that are gases at room temperature (other than the Group 0 elements) exist as pairs. For example, oxygen is O_2.
Symbol equations must be balanced. However, you can only change the numbers *in front* of each formula.
You may be asked to provide state symbols (s, l, g, aq).

3. Magnesium reacts with oxygen when it is heated. Complete the word and symbol equations to show this reaction. Include state symbols.

magnesium + oxygen → ____________________

2 ____(s) + ______(____) → ____MgO (s) [4 marks]

Scientific models of the atom

1. The diagram shows a model of the atom called the nuclear model.
Name the structures labelled X, Y and Z.

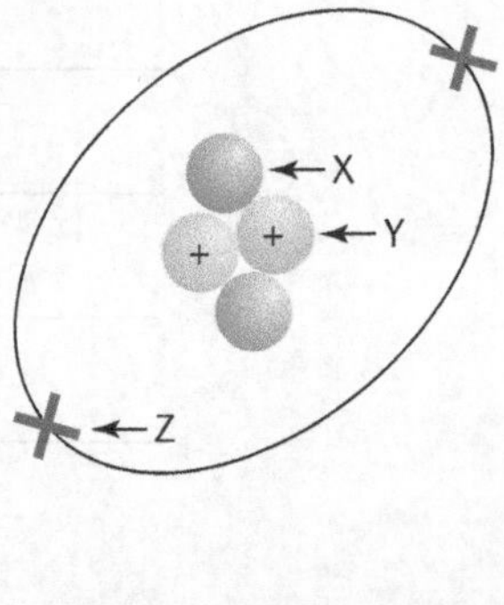

X ____________________

Y ____________________

Z ____________________ [3 marks]

2. An earlier model of the atom was the 'plum pudding' model. Compare the plum pudding model with the nuclear model shown on the previous page.

Command words

When you are asked to 'compare' things you need to write about the similarities and/or differences *between* them, not just write about one of the things.

[6 marks]

3. Literacy

Explain why the nuclear model of the atom is now accepted, but the plum pudding model is not.

Literacy

When you are asked to 'explain' something, write down the reasons why it happens.

You will get marks for how clear your answer is as well as for using key terms correctly.

[5 marks]

Sizes of atoms and molecules

1. Which of the following statements are correct?

Tick **two** boxes.

- [ ] All atoms have a nucleus.
- [ ] Atoms are the smallest particles that exist.
- [ ] Most of the mass of an atom is in the nucleus.
- [ ] The electrons, protons and neutrons all have the same mass.

[2 marks]

2. **Maths**

a Calculate the value of 3.5 m in mm.

Maths
In this question you will be converting numbers from one unit to another.

__________________ mm [2 marks]

Maths
1 km = 1000 m
1 m = 1000 mm (milli-)
1 mm = 1000 µm (micro-)
1 µm = 1000 nm (nano-)

b Calculate the value of 2 m in µm. Give your answer in standard form. [2 marks]

3. **Maths** The average radius of an atom is about 0.1 nm. (1 nm = 1×10^{-9} m.)

a Give the radius of an atom in m. Write your answer in standard form.

__________________ m [2 marks]

b The radius of a nucleus is less than 1/10 000 of that of an atom.

Calculate the radius of a nucleus in nm. Write your answer in standard form.

__________________ nm [1 mark]

Relative masses and charges of subatomic particles

1. Complete the table to show the relative mass and relative charge on the particles in an atom. [4 marks]

Particle	Relative mass	Relative charge
Electron		
Neutron	1	
Proton		+1

Command words
When you are asked to 'complete' something, answers should be written in the space provided.

2. Here is a diagram of an atom.

a Explain why an atom has no overall charge.

Remember

For an atom:

- its atomic number is the number of protons
- its mass number is the sum of the number of protons *and* neutrons.

____

____

____________________ [2 marks]

b Give the mass number of the atom. ____________ [1 mark]

c Give the atomic number of the atom. ____________ [1 mark]

d Name the element. ____________ [1 mark]

3. The diagram shows two different atoms.

Explain why these two atoms are isotopes of the same element.

____

____________________ [2 marks]

Relative atomic mass

1. There are two isotopes of bromine, $^{79}_{35}Br$ and $^{81}_{35}Br$.

a Give the atomic number of both isotopes. ____________ [1 mark]

b Calculate the number of neutrons in each isotope.

$^{79}_{35}Br$ ____________ $^{81}_{35}Br$ ____________ [2 marks]

c The relative atomic mass of bromine is 80.

Explain what this tells you about the abundances of each isotope in the sample.

Common misconception

The relative atomic mass is an average value that takes account of the abundance of the isotopes of the element.

____

____

______________________________ [2 marks]

2. In any sample of boron, 20% would be $^{10}_{5}B$ atoms and 80% would be $^{11}_{5}B$ atoms.

Maths Calculate the relative atomic mass of boron. Give your answer to 3 significant figures. [3 marks]

Worked Example Multiply the mass number of each isotope by the relative abundance (as the percentage).

$^{10}_{5}B = 10 \times 20 = 200$ $^{11}_{5}B = 11 \times 80 = 880$ [1 mark]

Add these numbers together and divide by 100: $\frac{200 + 880}{100}$ [1 mark] = 10.8

The relative atomic mass of boron is 10.8. [1 mark]

3. The relative abundance of silicon isotopes is 92% $^{28}_{14}Si$, 5% $^{29}_{14}Si$ and 3% $^{30}_{14}Si$.

Maths Calculate the relative atomic mass of silicon. Give your answer to 3 significant figures.

Relative atomic mass of silicon = ______________ [3 marks]

Electronic structure

1. Sodium has the atomic number 11.

a Draw a diagram to show the electronic structure of a sodium atom.

Analysing questions

The atomic number of the element gives the number of electrons. You can use the periodic table to find out the atomic number from the element name.

Command words

When you are asked to 'draw', you need to produce or complete a diagram.

[3 marks]

b When sodium reacts with non-metals it loses one electron to form a charged particle called an ion. Give the charge on a sodium ion and explain why it has this charge.

____

____

____

____ [3 marks]

2. The diagram shows an oxide ion.

a Write the electronic structure of the oxide ion.

____ [1 mark]

b Give the charge of the oxide ion.

____ [1 mark]

Electronic structure and the periodic table

1. The periodic table is used to organise the known chemical elements.

Use words from the box to complete the sentences.

atomic	chemical	group	ionic	period	section

Elements are arranged in the periodic table in order of their ____________________ number.

Elements with similar properties are found in the same ____________________. [2 marks]

2. The diagram shows the periodic table. Some elements are shown by letters. These letters are **not** the usual chemical symbols.

[diagram: periodic table with group labels 2 and 3 and elements W, Y, X, Z marked]

a Give the number of electrons atoms of element X will have in their outer shell.

______________ [1 mark]

b Which element shown in the diagram will have the highest atomic number?

Tick **one** box.

☐ W ☐ X ☐ Y ☐ Z [1 mark]

c Explain why element Z is unreactive.

____

______________________________ [2 marks]

d Element W is added to water. Bubbles form slowly. If element Y is added to water, you can predict that bubbles will also form. Explain why you can make this prediction.

____

______________________________ [2 marks]

Development of the periodic table

1. In 1869 Dmitri Mendeleev produced an early version of the periodic table. He arranged the elements in rows in order of atomic weight. The elements in the modern periodic table are arranged in order of atomic number. Explain why Mendeleev could not use atomic number in his table.

____

______________________________ [2 marks]

2. The table shows part of Mendeleev's periodic table.

Group 1	Group 2	Group 3	Group 4	Group 5	Group 6	Group 7
H						
Li	Be	B	C	N	O	F
Na	Mg	Al	Si	P	S	Cl
K Cu	Ca Zn	# #	Ti #	V As	Cr Se	Mn Br
Rb Ag	Sr Cd	Y In	Zr Sn	Nb Sb	Mo Te	# I

a Mendeleev left gaps in his periodic table, which are shown by '#'. Explain why he decided to leave gaps.

____

____

________________________________ [2 marks]

b Mendeleev placed iodine(I), which has an atomic weight of 127, after tellurium (Te) even though tellurium has an atomic weight of 128. Suggest why he did this.

____

________________________________ [1 mark]

3. At the time many scientists did not accept Mendeleev's periodic table. Suggest two problems with his arrangement of elements in Group 1.

Analysing questions

Compare the modern periodic table with Mendeleev's periodic table.

You know that elements in a group should share similar properties.

Look at the elements in Group 1 of Mendeleev's periodic table. Do they share similar properties?

____

________________________________ [2 marks]

Comparing metals and non-metals

1. Which statements are describing metals?

Tick **two** boxes.

☐ Form negative ions ☐ Found on the left side of the periodic table

☐ Form positive ions ☐ Found on the right side of the periodic table [2 marks]

2. Element X reacts with chlorine to form a compound with the formula XCl_3.

a For the two elements below, identify if it is a metal or non-metal.

Element X ____________________

Chlorine ____________________ [2 marks]

b Predict which group in the periodic table element X is in. Give a reason for your answer.

____

____ [2 marks]

3. Read the information about the physical properties of two elements.

Potassium is a Group 1 element. It is a shiny material that conducts electricity and floats on water. Graphite is a form of carbon (a Group 4 element). It has a dull appearance, is solid at room temperature and conducts electricity.

For each element describe the property which is not typical. Give a reason for each answer.

Potassium ____

____

Graphite ____

____ [4 marks]

Elements in Group 0

1. Which statement is true about **all** the elements in Group 0?

Tick **one** box.

☐ They exist as molecules.

☐ They exist as single atoms.

☐ They have eight electrons on their outer shell.

☐ They have high melting and boiling points. [1 mark]

Remember

Helium is in Group 0 but only has two electrons on its outer shell.

Analysing questions

Only a few words will be in **bold** in a question paper. Make sure you take particular note of them.

2. The diagram shows an atom of a Group 0 element.

Synoptic

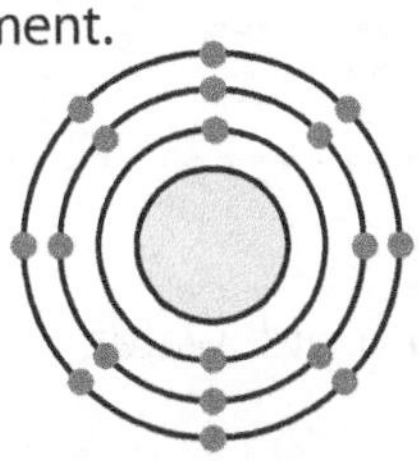

a Name the element shown in the diagram. ________________ [1 mark]

Remember

Use the periodic table. Counting the number of electron shells will give you the period the element is in.

b Use the electronic structure of the element shown in the diagram to explain why it is unreactive.

____

____

____ [3 marks]

3. The table shows some data on the Group 0 elements.

Element	Atomic number	Boiling point (°C)
Helium	2	−268.9
Neon	10	−246.1
Argon	18	−185.8
Krypton	36	−153.22
Xenon	54	−108.1

a Describe the trend shown by the data in the table.

____ [1 mark]

Synoptic

b Mixtures of gases can be separated by cooling the mixture. Describe how a mixture of helium and neon can be separated using this method.

____

____

____ [3 marks]

Elements in Group 1

1. A teacher reacts sodium with water.

Required practical

a Complete the **balanced** symbol equation to show this reaction.

2____ + $2H_2O$ → ____NaOH + ____ [3 marks]

b Describe **one** observation the students would see during the reaction.

____

______________________________ [1 mark]

2. Use information from the table to answer the following questions.

Element	Atomic number	Melting point (°C)
Lithium	3	181
Sodium	11	98
Potassium	19	64
Rubidium	37	–
Caesium	55	29

Maths

a Plot the data on the graph axes below.
Draw a line of best fit. [3 marks]

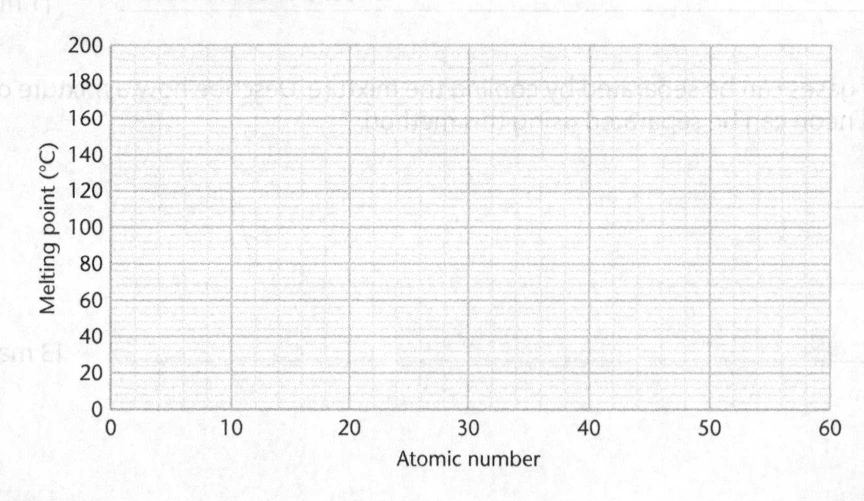

Maths
You can use your line of best fit to estimate values for missing data.

Maths

b Estimate the melting point of rubidium.

______________ [1 mark]

Command words
'Estimate' means give an approximate answer.

Elements in Group 7

1. Synoptic

Give the relative molecular mass of chlorine **molecules**.

You need to use the periodic table to find the data for your answer.

______________ [2 marks]

2. The elements in Group 7 all have similar reactions.

a Explain, in terms of their electronic structures before and after a reaction, why Group 7 elements have similar reactions.

____

____ [2 marks]

b Complete the balanced symbol equation for the reaction of bromine with potassium.

2K(s) + _____(l) → 2_____(s) [2 marks]

c Describe how the reactivity of the Group 7 elements changes as you go down the group.

____ [1 mark]

Analysing questions

These are the steps to use when you are asked to write an ionic equation:

1. Write out the full balanced symbol equation. For example:

   $Cl_2(g) + 2KBr(aq) \rightarrow 2KCl(aq) + Br_2(aq)$

2. Add charges to the ions.

   $Cl_2(g) + 2K^+ 2Br^-(aq) \rightarrow 2K^+ 2Cl^-(aq) + Br_2(aq)$

3. Cross out any ions that are not involved in the reaction.

   $Cl_2(g) + \cancel{2K^+} 2Br^-(aq) \rightarrow \cancel{2K^+} 2Cl^-(aq) + Br_2(aq)$

4. The ionic equation is the material that remains.

   $Cl_2(g) + 2Br^-(aq) \rightarrow 2Cl^-(aq) + Br_2(aq)$

Command word

Note that the command word here is 'describe' and not 'explain'.

3. A student bubbled chlorine gas through a solution of colourless sodium bromide solution.

Higher Tier only

a Give an ionic equation for the reaction.

______________________________ [2 marks]

b Describe what the student would see. Explain your observations.

Command words

In this question you are being asked to describe the appearance before *and* after the reaction.

____

____

______________________________ [3 marks]

Properties of the transition metals

1. A compound has the formula $NiSO_4$.

a Name the transition metal in the compound. ______________ [1 mark]

b Determine the charge of the transition metal ion. Explain your answer.

______________________________ [2 marks]

c Suggest one use for this compound.

______________________________ [1 mark]

2. The transition elements are a group of metals with similar physical properties.

Remember

Physical properties are properties such as hardness, strength, size, shape, melting and boiling points. They can be measured without carrying out a chemical reaction.

Chemical properties relate to how a substance reacts.

a Copper and gold are both transition metals. Give **two** physical properties shown by both copper and gold.

1. ______________________________

____

2. ______________________________ [2 marks]

b The elements in Group 1 are also metals. Describe a difference between the chemical properties of Group 1 metals and transition metals.

______________________________ [1 mark]

3. Bone implants are used to replace or strengthen damaged bones. They are usually made from the transition metal titanium because it has a relatively low density.

a Give **one** other physical property that makes titanium a good choice for a bone implant.

______________________________ [1 mark]

b The Group 1 metal lithium has a much lower density than titanium. Explain why lithium is not used for bone implants. Use the chemical and physical properties of lithium in your answer.

____

____

____

______________________________ [4 marks]

Analysing questions

To get full marks describe one chemical property and one physical property of lithium. For each property, explain why this means that lithium is not suitable for this function.

The three states of matter

1. The table shows the melting points and boiling points of three substances.

Substance	Melting point (°C)	Boiling point (°C)
X	−97.6	65
Y	−259	−253
Z	801	1413

a Which substance is a liquid at room temperature (25 °C)?
Tick **one** box.

☐ X ☐ Y ☐ Z [1 mark]

b Give the state of Z at a temperature of 700 °C.

_______________ [1 mark]

c Draw particles in the box below to represent Z in this state.

[1 mark]

Higher Tier only

d Describe the arguments for and against using the model you used in part c to show the structure of Z.

____

____

____

____

____

_______________ [6 marks]

Ionic bonding and ionic compounds

1. The ionic compound calcium nitrate is made up of Ca^{2+} ions and NO_3^- ions.

What is the formula of calcium nitrate?

Tick **one** box.

☐ $CaNO_3$ ☐ $Ca(NO_3)_2$ ☐ Ca_2NO_3 ☐ $Ca_2(NO_3)_2$ [1 mark]

2. **a** Give the formula of calcium chloride. [1 mark]

Worked Example

Calcium is in Group 2 of the periodic table, so its ions have a charge of +2.

Chlorine is in Group 7, so its ions have a charge of −1.

The charges on the positive ions must equal the charges on the negative ions.

The formula of calcium chloride is $CaCl_2$. [1 mark]

b The diagram shows two ball-and-stick diagrams.

A

B

Which diagram represents calcium chloride?

_______________ [1 mark]

Remember

Each sphere represents a type of ion. You can count them to give you an estimate of the ratio.

It will not be an exact number because the diagrams show just part of a much bigger structure. The ions on the sides and corners are bonded to other ions that are not included in the diagram.

c Explain your answer in part b.

____

____

____ [2 marks]

Dot and cross diagrams for ionic compounds

1. The diagram shows an atom of the Group 1 element sodium.

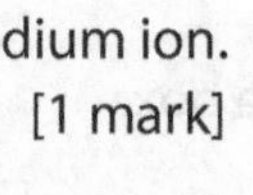

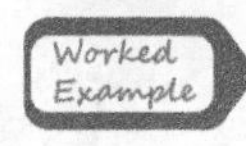

a Describe what must happen to turn a sodium atom into a sodium ion. [1 mark]

It loses the outer electron. [1 mark]

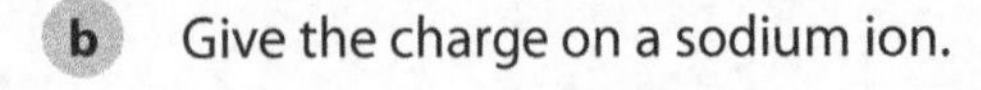

b Give the charge on a sodium ion.

______________________ [1 mark]

> **Remember**
> An atom has no overall charge because it has equal numbers of positive protons and negative electrons. If an atom gains or loses electrons its charge will change.

c Explain why a sodium ion has this charge.

____ [1 mark]

2. Draw a dot and cross diagram to show what happens when sodium reacts with chlorine. Give the name of the compound and its formula.

[5 marks]

Properties of ionic compounds

1. A teacher set up the equipment as represented by the diagram.

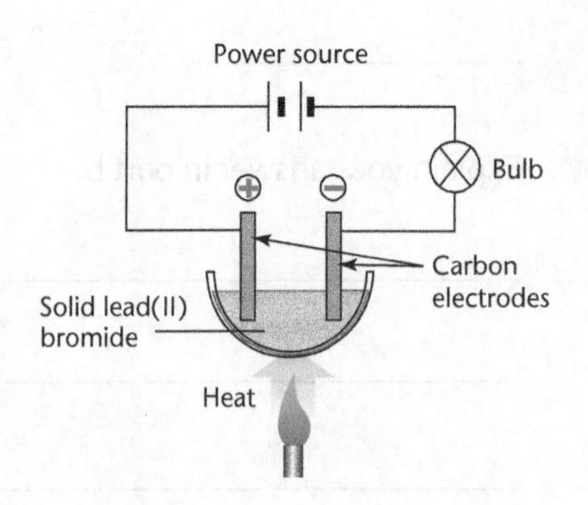

Synoptic **a** Give the formula of lead(II) bromide. ______________ [1 mark]

b At the start of the experiment the bulb did not light up. However, as the experiment continued the bulb did light up. Explain these observations.

____

____

____ [3 marks]

c The melting point of lead(II) bromide is 370.6 °C. Explain why lead(II) bromide has a high melting point. You should describe its structure in your answer.

____

____

____ [3 marks]

Covalent bonding in small molecules

1. Use formulae from the box to answer the questions.

Synoptic

Ca $CaCl_2$ Cl_2 CO_2

a Which substance is a compound formed of small molecules?

______________ [1 mark]

b Which substance is an ionic compound?

______________ [1 mark]

c Which substance is an element formed of small molecules?

______________ [1 mark]

d Which **two** substances contain covalent bonds?

____ [2 marks]

2. Ethane has the formula C_2H_6.

Complete the diagram, drawing **lines** to show the bonds in ethane.

H H

H C C H

H H

[2 marks]

Dot and cross diagrams for covalent compounds

1. The diagram shows how the outer electrons are arranged in an atom of hydrogen and an atom of chlorine.

Hydrogen atom

Chlorine atom

Cl

Draw a diagram to show how the atoms and outer electrons are arranged in a molecule of hydrogen chloride.

[3 marks]

Each atom needs a full outer shell of electrons. For hydrogen, this is two electrons (it only needs to fill the first shell) and for chlorine this is eight electrons.

By sharing one electron each, both atoms achieve a full outer shell.

To show this, overlap the outer shells of electrons and place the shared pair (or pairs) where the outer shells overlap.

2. Draw a dot and cross diagram to show the covalent bonding in a molecule of oxygen (O_2). Show the outer electron shells only. Use the space below.

Remember

More than one pair of electrons can be shared.

[3 marks]

Properties of small molecules

1. Substance Z is made up of small molecules.

Synoptic The diagram shows how two molecules of Z are arranged when the substance is a liquid.

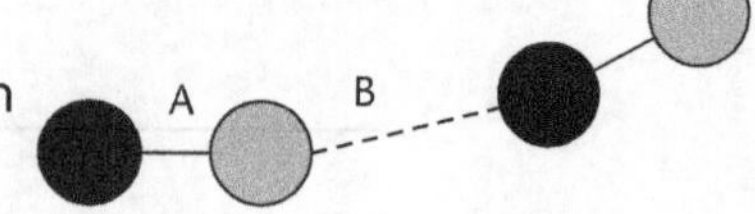

a Name the force or bond labelled A.

____

[1 mark]

b Which bond or force is overcome when substance Z boils?

______________________________ [1 mark]

> **Common misconception**
> It is important to remember that there are strong bonds between the atoms in a molecule but weak forces between the molecules.

2. Methane consists of small molecules. Methane is a gas at 25 °C.

Synoptic **a** Describe what this tells you about its boiling point.

______________________________ [1 mark]

b Explain, using what you know about bonding, why methane is a gas at room temperature.

____

______________________________ [2 marks]

Polymers

1. The diagram shows two polymer chains.

a Draw **one** intermolecular force. [1 mark]

b Draw an arrow to mark a region where there are strong covalent bonds. [1 mark]

> **Remember**
> **Inter**molecular means between molecules.
> **Intra**molecular means between the atoms in molecules.

2. Poly(ethene) is a polymer of ethene.

Poly(ethene) can be shown as:

$$\left(\begin{array}{c} H \quad H \\ | \quad\; | \\ -C-C- \\ | \quad\; | \\ H \quad H \end{array} \right)_n$$

Describe what the '*n*' shows.

______________________________ [1 mark]

Giant covalent structures

1. The diagrams show part of the structure of three giant covalent structures.

a Name each giant covalent structure.

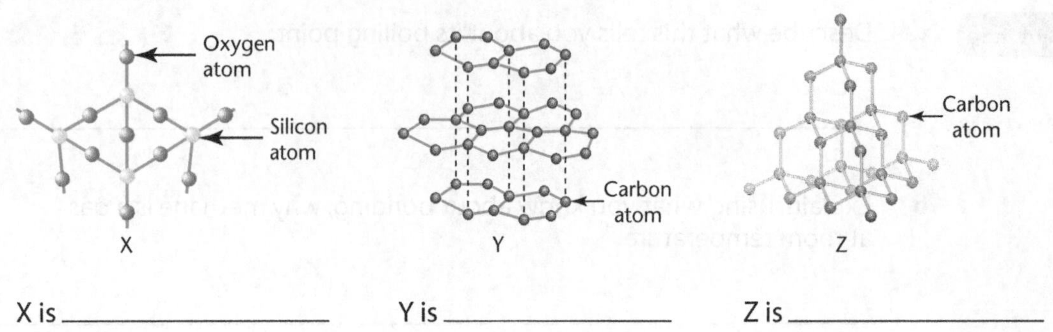

X is ______________ Y is ______________ Z is ______________

[3 marks]

Synoptic **b** Name the structure that shows a compound. ______________ [1 mark]

2. Silicon dioxide (SiO_2) and carbon dioxide (CO_2) are both covalent compounds. Compare the structures of silicon dioxide and carbon dioxide. You should describe the bonding between the atoms in the two compounds.

____

____

____

____ [4 marks]

Properties of giant covalent structures

1. Pure quartz is silicon dioxide. It has a giant covalent structure.

Synoptic **a** Define what is meant by a 'pure' substance.

____ [1 mark]

b Pure quartz has a high melting point. Use what you know about the structure of quartz to explain why.

____

____ [2 marks]

2. The diagram shows the structures of diamond and graphite.

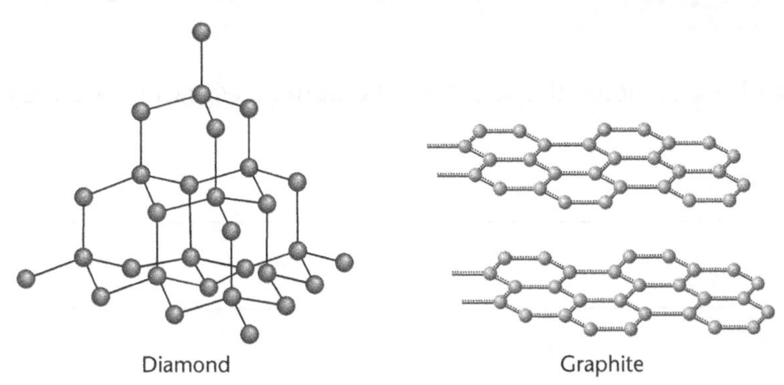

Command words

When you are asked to 'compare' you need to describe the similarities and/or differences between things, not just write about one of them.

Compare the physical properties of diamond and graphite, linking their properties to their chemical structures.

____

____

____

____

____

____ [6 marks]

Graphene and fullerenes

1. The diagram shows part of a carbon nanotube.

Maths

Graphene is a form of carbon. Compare the structure of graphene with the structure of graphite.

____

____

____

______________________________ [4 marks]

Nanoparticles

1. Titanium dioxide particles are used in sunscreen. They have a diameter of 10 nm.

a Explain why titanium dioxide particles are classed as nanoparticles.

______________________________ [1 mark]

b A water molecule has a diameter of 0.1 nm.

Calculate the orders of magnitude between the diameter of a water molecule and the diameter of a titanium dioxide particle.

____

______________________________ [2 marks]

2. Gold can be used as a catalyst.

Maths

The cube of gold shown has sides that are 2 cm in length.

2 cm

Synoptic **a** Define 'catalyst'.

____

____ [2 marks]

b Calculate the surface area to volume ratio of this cube of gold. Show your working.

________________ [3 marks]

c Explain why the gold would be a more effective catalyst if it was cut up into smaller pieces.

____

____

____ [3 marks]

Uses of nanoparticles

1. Literacy

$PM_{2.5}$ particulates are produced when fuels burn. A scientific study carried out in 2011 estimated that in the UK 90 deaths per year are caused by people breathing in $PM_{2.5}$ particulates.

Evidence like this has prompted people to think that the use of nanoparticles in products such as sunscreen should be banned.

Use the information given above, plus your own knowledge on the benefits and risks of nanoparticles, to evaluate a ban on nanoparticles.

Command words

When you are asked to 'evaluate' something you should use the information supplied as well as your knowledge and understanding to consider evidence for and against.

In this question you should describe the arguments for and against using nanoparticles.

____

____

____

____

____

____ [6 marks]

Metallic bonding

1. Which statements describe the atoms in metals?

Tick **two** boxes.

- ☐ They are arranged in a regular pattern.
- ☐ They are bonded ionically.
- ☐ They are attracted by strong bonds.
- ☐ They do not have delocalised electrons.

[2 marks]

2. The diagram shows metallic bonding.

Label the two items in the diagram.

[2 marks]

3. The electrons in a metal are delocalised. Explain the meaning of 'delocalised'.

____

______________________________ [2 marks]

4. A student suggests the following hypothesis: 'The more delocalised electrons in the structure of a metal, the higher its melting point.'

a Calcium has a much higher melting point than sodium.

Describe how this information supports the hypothesis.

Remember

A hypothesis is a possible explanation for an observation.

____

____

______________________________ [2 marks]

b Suggest what the student could do to strengthen the hypothesis.

____

____

____ [3 marks]

Properties of metals and alloys

1. The diagram shows the structures of two different metals.

A B

a Define the term 'alloy'.

____ [1 mark]

b Which of the metals in the diagram is an alloy?

Tick **one** box.

☐ A ☐ B [1 mark]

c Explain your answer to part b.

____ [1 mark]

d Use the diagrams to explain why metals are malleable.

____

____ [2 marks]

e Use the diagrams to explain why alloys are stronger than pure metals.

____

____ [2 marks]

Writing formulae

1. Write the formulae of the following compounds.

> **Remember**
> Here are the formulae of some common ions:
> OH^- NO_3^- CO_3^{2-} SO_4^{2-}
> The charges in a compound must balance.

Synoptic **a** Sodium sulfate

Worked Example Na_2SO_4 [1 mark]

b Copper(II) carbonate

______________ [1 mark]

c Magnesium nitrate

______________ [1 mark]

d Calcium hydroxide

______________ [1 mark]

2. Give the number of hydrogen atoms in each of the compounds below.

a CH_4

____

b (structure: C bonded to OH, H_3C, CH_3 and H)

____

c $CH_3(CH_2)_2COOH$

______________ [3 marks]

3. A scientist carries out the following reaction:

$CH_4 + 4Cl_2 \rightarrow CCl_4 + 4HCl$

Synoptic **a** Name the product that is an acid.

______________ [1 mark]

Synoptic **b** Suggest a method the scientist could use to show that an acid was formed in the reaction.

____

______________ [2 marks]

Conservation of mass and balanced chemical equations

1. A student uses the equipment below to heat magnesium. The reaction forms solid magnesium oxide.

Required practical

This is the method the student follows.

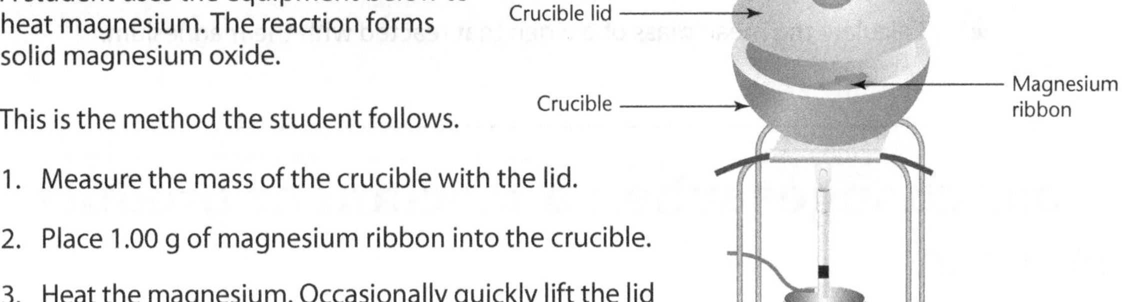

1. Measure the mass of the crucible with the lid.
2. Place 1.00 g of magnesium ribbon into the crucible.
3. Heat the magnesium. Occasionally quickly lift the lid up using tongs and place it back down.
4. When you see no further reaction leave the crucible to cool.
5. Measure the mass of the crucible, the lid and the contents.

Remember

When you balance an equation you cannot change the small (subscript) numbers as this would change the formula of the substance. However, you can change the number of atoms of each substance by adding a number *in front* of the formula.

a Write the balanced symbol equation for this reaction. Include state symbols.

____ [3 marks]

b Describe how the student could calculate the mass of magnesium oxide formed.

____ [1 mark]

Maths

c The student repeats the method two more times. Each time they calculate the mass of magnesium oxide formed:

Trial 1 = 1.48 g; Trial 2 = 1.38 g; Trial 3 = 1.50 g

Calculate the mean mass of magnesium oxide formed. Give the uncertainty in your answer.

Maths

The uncertainty is the range of a set of measurements about the mean. Range = the highest measurement – the lowest measurement, so uncertainty = ± half the range.

mean mass = _____ g ± _____ g [3 marks]

d The mass of magnesium oxide produced was lower than expected. Suggest **two** experimental errors that may have caused this.

____

____ [2 marks]

e Calculate the mean mass of oxygen that reacted with the magnesium.

____ [1 mark]

Mass changes when a reactant or product is a gas

1. Required practical

A student investigated how the mass changed during the reaction between magnesium carbonate and hydrochloric acid.

The balanced symbol equation for the reaction is:

$MgCO_3(s) + 2HCl(aq) \rightarrow MgCl_2(aq) + H_2O(l) + CO_2(g)$

The diagram shows the equipment they used.

a The function of the cotton wool bung is to prevent any liquid escaping.

Suggest a reason why a normal rubber bung is not used.

____

____

____ [3 marks]

b Predict how the mass recorded on the balance would change during the experiment.

Give a reason for your answer.

Command words

When you are asked to 'predict' you should give a plausible outcome.

____

____

____ [3 marks]

Synoptic **c** After 4 minutes the student observed that there was still some magnesium carbonate in the conical flask but the mass stopped changing.

Explain this observation.

____ [1 mark]

d The student repeated the experiment and measured the mass change a further four times.
Describe how the student should calculate the uncertainty in the results.

____ [1 mark]

Relative formula mass

1. How do you calculate the relative formula mass (M_r) of a compound?

Tick **one** box.

- [ ] Add up the atomic numbers of the atoms.
- [ ] Add up the relative atomic masses of the atoms.
- [ ] Count the number of atoms.
- [ ] Measure the mass of the compound. [1 mark]

2. Give the relative formula mass, M_r, of copper(II) nitrate, $Cu(NO_3)_2$. [1 mark]

Maths — Relative atomic masses, A_r: N = 14; O = 16; Cu = 63.5

Worked Example

A_r Cu = 63.5 M_r $NO_3 = 14 + (16 \times 3) = 62$

M_r $Cu(NO_3)_2 = 63.5 + (2 \times 62) = 187.5$ [1 mark]

Remember
If the relative atomic masses (A_r) of elements are not given in the exam question, you can use your copy of the periodic table.

3. The relative formula mass, M_r, of a Group 2 metal sulfate is 136.

Maths — Calculate the relative atomic mass, A_r, of the Group 2 metal and so identify the metal.

Synoptic — Relative atomic masses, A_r: O = 16; S = 32

Relative atomic mass, A_r = ____________ Metal is ____________ [4 marks]

4. The equation shows a reaction:

Maths

$X + Y \rightarrow C_4H_8O_2 + Z$ M_r: X = 46; Y = 60

Calculate the M_r of Z.

[3 marks]

Moles

1. There are 6.02×10^{23} atoms in 1 mole of hydrogen atoms.

Higher Tier only

Maths

How many molecules are there in 1 mole of methane, CH_4?

Tick **one** box.

☐ 3.01×10^{23} ☐ 6.02×10^{23} ☐ 1.20×10^{24} ☐ 2.41×10^{24} [1 mark]

2. Calculate the mass in grams of the following quantities.

Maths

> **Remember**
>
> You can calculate mass of a substance by using the formula:
>
> mass = moles × relative atomic mass (A_r) or
>
> mass = relative formula mass (M_r)
>
> This equation can be rearranged:
>
> $\text{moles} = \frac{\text{mass}}{A_r}$ or $\frac{\text{mass}}{M_r}$

Worked Example

a 2 moles of carbon.

The A_r for a carbon atom is 12.

mass = moles × A_r

mass of C = 2 × 12 = 24 g [1 mark]

b 0.5 moles of sodium chloride.

Relative atomic masses, A_r: Na = 23; Cl = 35.5

[2 marks]

3. Calculate the number of moles in 283.5 g of hydrogen bromide (HBr).

Maths

[2 marks]

4. 2.5 moles of element X has a mass of 17.5 g.

Maths

Calculate the atomic mass of element X. Show the calculation you used.

[1 mark]

5. Propane is a fuel. When propane is burnt in sufficient oxygen the following chemical reaction takes place:

$$C_3H_8 + 5O_2 \rightarrow 3CO_2 + 4H_2O$$

a How many moles of oxygen **molecules** are used? ________________ [1 mark]

b How many moles of oxygen **atoms** are produced? ________________ [1 mark]

Amounts of substances in equations

1. Aluminium is heated in oxygen.

Higher Tier only

The balanced symbol equation for the reaction is:

$4Al + 3O_2 \rightarrow 2Al_2O_3$

How many moles of aluminium oxide are produced if 6 moles of aluminium are used?

Tick **one** box.

☐ 2 ☐ 3 ☐ 4 ☐ 6 [1 mark]

2. A student reacted 1.2 g of magnesium with excess hydrochloric acid.

The balanced symbol equation for the reaction is:

$Mg(s) + 2HCl(aq) \rightarrow MgCl_2(aq) + H_2(g)$

Synoptic

a Describe a test the student can carry out to prove the gas made was hydrogen.

____

____ [2 marks]

Maths **b** Calculate the maximum mass of magnesium chloride that can be made.
Show your working.

Analysing questions

Follow these stages:

1. Calculate the moles of magnesium.
2. The symbol equation shows that the number of moles of magnesium = moles of magnesium chloride
3. Calculate the mass of magnesium chloride, using the number of moles you have just found and the relative atomic masses for magnesium and chlorine.

________________ g [3 marks]

Using moles to balance equations

1. 127 g of copper is heated in oxygen to produce 159 g of copper(II) oxide.

Higher Tier only **a** Calculate the number of moles of copper.

Maths

[1 mark]

Worked Example **b** Show how you can carry out a calculation to prove that the balanced symbol equation is

$2Cu + O_2 \rightarrow 2CuO$ [2 marks]

There are 2 moles of copper.

moles of copper(II) oxide = $\frac{159}{63.5 + 16}$ = 2 moles [1 mark]

The number of moles of copper and of copper oxide are equal, which is what is shown in the balanced symbol equation. [1 mark]

2. 4.6 g of sodium reacts with dilute hydrochloric acid to form 11.7 g of sodium chloride.

Synoptic **a** Name the other product formed in this reaction.

______________________________ [1 mark]

Maths **b** Calculate the ratio of moles of Na:NaCl.

Relative atomic masses, A_r: Na = 23; Cl = 35.5

Maths

A ratio shows the relative sizes of two or more values.

So, if a ratio is 1A:2B then this means there are twice as many B as A.

____

____

____

____

______________________________ [3 marks]

Limiting reactants

1. A student adds some pieces of magnesium ribbon to a test tube of dilute hydrochloric acid.

Higher Tier only

The student observes fizzing. After 9 seconds the fizzing stops and the magnesium appears to have disappeared.

Name the limiting reactant. Give a reason for your answer.

____

____

______________________________ [2 marks]

2. The thermite reaction is used to produce molten iron.

The balanced symbol equation for the reaction is:

$Fe_2O_3 + 2Al \rightarrow 2Fe + Al_2O_3$

100 g of iron oxide is reacted with 100 g of aluminium.

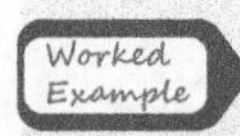

a Calculate which reactant is the limiting factor in the reaction.

Relative atomic masses, A_r: O = 16; Al = 27; Fe = 56 [4 marks]

moles of $Fe_2O_3 = \frac{100}{(56 \times 2) + (16 \times 3)} = 0.625$ [1 mark]

From the equation, the ratio is 1 mole Fe_2O_3 to 2 moles Al.

$0.625 \times 2 = 1.25$ moles of aluminium will be needed to react exactly with the Fe_2O_3. [1 mark]

moles of Al present $= \frac{100}{27} = 3.70$ [1 mark]

Aluminium is present in excess, so Fe_2O_3 is the limiting factor. [1 mark]

b Calculate the maximum mass of iron that can be produced.

[2 marks]

3. 11.5 g of sodium is added to 15 g of chlorine.

The reaction that takes place is:

$2Na + Cl_2 \rightarrow 2NaCl$

Show which reactant is the limiting factor in the reaction.

[4 marks]

Concentration of solutions

1. Calculate the mass of sodium chloride needed to make 100 cm^3 of a 15 g/dm^3 solution.

Maths

Maths

The units of concentration show you how to calculate concentration.

If the unit is g/cm^3 then you need to divide the mass of the solute (in g) by the volume of the solvent (in cm^3). You may have to convert units to do this, for example from dm^3 to cm^3.

The formula is:

$$\text{concentration (g/cm}^3\text{)} = \frac{\text{mass (g)}}{\text{volume (cm}^3\text{)}}$$

This formula can be rearranged:

mass = volume × concentration

$$\text{and volume} = \frac{\text{mass}}{\text{concentration}}$$

______________ g [2 marks]

2. A student dissolves 6.2 g of copper sulfate crystals into 0.05 dm^3 of water to produce a pale blue solution.

Maths

a Calculate the concentration of the solution in g/cm^3.

Give your answer to 2 significant figures.

______________ g/cm^3 [2 marks]

Higher Tier only

b The student pours the solution into a beaker and leaves it uncovered in the classroom. After 3 days the colour of the solution is much darker. Suggest why.

Analysing questions

Think about what happens to the water in the solution when it is left, and how this will affect the concentration of copper sulfate solution.

____

____

____

____

____ [3 marks]

3. Calculate the concentration (in g/cm³) of the solution made when 0.25 moles of potassium iodide (KI) are dissolved in 250 cm³ of water.

Maths

Higher Tier only

Synoptic

Relative atomic masses, A_r: K = 19; I = 53

_______________ g/cm³ [3 marks]

Using concentrations of solutions in mol/dm³

1. A scientist made a solution of copper(II) chloride ($CuCl_2$). The scientist dissolved 269 g of solid in 2 dm³ of water.

Maths

Higher Tier only

a Calculate the concentration of the solution in mol/dm³.

_______________ mol/dm³ [3 marks]

b The scientist used 50 cm³ of the solution in an experiment.

Calculate the mass of copper(II) chloride in this volume of solution. Give your answer to 1 decimal place.

_______________ g [3 marks]

Amounts of substances in volumes of gases

1. Maths | Higher Tier only

Calculate the volume occupied by 220 g of carbon dioxide at room temperature and pressure.

_______________ dm^3 [2 marks]

2. A student placed hydrochloric acid and zinc carbonate into a flask.

The reaction that took place is shown by the symbol equation:

$ZnCO_3(s) + 2HCl(aq) \rightarrow ZnCl_2(aq) + H_2O(l) + CO_2(g)$

The student measured the mass of the flask and reactants at the start of the reaction and again when the reaction was over. The table shows the results.

Remember

At room temperature and pressure (20 °C and 1 atm) the volume of 1 mole of **any** gas is 24 dm^3. You will need to remember this for the exam.

Mass at start in g	Mass at end in g
256.50	256.05

Synoptic

a Describe how the student could tell when the reaction was over.

_______________ [1 mark]

Maths | Higher Tier only

b Calculate the volume of the carbon dioxide produced in the reaction. Give your answer to 3 significant figures.

_______________ dm^3 [3 marks]

Maths

When using a calculator to carry out a multi-step calculation, do not round up numbers between each stage. Use the full number on the calculator in the next step of the calculation. Then use the instructions in the question at the end of the calculation.

Percentage yield

1. 16 g of iron is produced in a reaction. The maximum theoretical mass of iron that could be produced was 32 g.

Worked Example

Calculate the percentage yield in the reaction. [2 marks]

$$\% \text{ yield} = \frac{\text{mass of product actually made}}{\text{maximum theoretical mass of product}} \times 100 \text{ [1 mark]}$$

$$\% \text{ yield} = \frac{16}{32} \times 100 = 50\% \text{ [1 mark]}$$

2. The Haber process is an industrial process used to make ammonia.

The Haber process reaction can be represented by this equation:

$N_2(g) + 3H_2(g) \rightleftharpoons 2NH_3(g)$

Higher Tier only

a Calculate the maximum mass of ammonia that can be made from 2000 g of nitrogen.
Relative atomic masses, A_r: H = 1; N = 14

[3 marks]

b At a temperature of 450 °C and 200 atmospheres the actual mass of ammonia produced from 2000 g of nitrogen is 608 g.
Calculate the percentage yield of ammonia.

[2 marks]

Atom economy

Remember

Atom economy is shown as a percentage and is calculated using this formula:

$$\frac{\text{relative formula mass of desired product from equation}}{\text{sum of relative formula masses of all reactants from equation}} \times 100$$

1. Magnesium nitrate is a salt that can be used in fertilisers.

Maths

A company manufactures magnesium nitrate using the reaction shown in this symbol equation:

$MgCO_3(s) + 2HNO_3(aq) \rightarrow Mg(NO_3)_2(aq) + H_2O(l) + CO_2(g)$

Relative formula masses, M_r: $MgCO_3 = 84$; $HNO_3 = 63$; $Mg(NO_3)_2 = 148$; $H_2O = 18$; $CO_2 = 44$

a Calculate the atom economy of this reaction.

[3 marks]

Another reaction can be used to make magnesium nitrate:

$MgO(s) + 2HNO_3(aq) \rightarrow Mg(NO_3)_2(aq) + H_2O(g)$

b Calculate the atom economy for this reaction.

[3 marks]

c Compare the atom economies for the two reactions. Give a reason for the difference.

____

____ [2 marks]

Higher Tier only

d Suggest why the company uses the first reaction despite the lower atom economy.

Command words

'Suggest' requires you to use your knowledge and understanding to think of a possible reason.

____ [1 mark]

Metal oxides

1. The symbol equation shows the reaction for a hydrocarbon burning in air:

$2C_2H_6 + 7O_2 \rightarrow$ ____ $CO_2 + 6H_2O$

a Write a number in the space to complete the balancing of the equation. [1 mark]

Synoptic **b** Name the hydrocarbon. ____________________ [1 mark]

c Name the two elements that are oxidised in this reaction.

____________________ ____________________ [2 marks]

2. Iron is produced in industry by the following reaction:

$Fe_2O_3 + 3CO \rightarrow 2Fe + 3CO_2$

a Name the substance that is reduced during this reaction.

____ [1 mark]

Higher Tier only | Synoptic | Maths

b Calculate the maximum mass of iron produced from 20 kg of iron oxide.

Relative atomic masses, A_r: O = 16; Fe = 56

Remember

$$\text{moles} = \frac{\text{mass}}{\text{relative formula mass}}$$

____________________ kg [3 marks]

Reactivity series

1. Which statement describes what happens when lithium reacts?

Tick **one** box.

☐ Lithium gains one electron.
☐ Lithium gains one proton.
☐ Lithium loses one electron.
☐ Lithium loses one proton.

[1 mark]

2. Explain why lithium is less reactive than sodium.

____

____

______________________________ [3 marks]

3. A student was given pieces of four different metals (A–D).

The student placed each piece in a different test tube of water. The student's observations are shown in the table.

Metal	Observation
A	A few bubbles were produced
B	Very vigorous bubbling
C	No bubbles
D	Vigorous bubbling

a Write the metals in order of reactivity in the space below, the **least** reactive first.

_______ _______ _______ _______ [1 mark]

b Give the name of a metal that could be metal C. ______________ [1 mark]

c Explain why the student should not add metal B to acid.

____

______________________________ [2 marks]

Reactivity series – displacement

1. Use the reactivity series to answer these questions.

Complete these displacement reaction equations.

Most reactive
Potassium
Sodium
Calcium
Magnesium
Aluminium
Carbon
Zinc
Iron
Tin
Lead
Hydrogen
Copper
Silver
Gold
Platinum
Least reactive

Worked Example

a magnesium carbonate + *sodium* [1 mark] → *magnesium* [1 mark] + sodium carbonate [2 marks]

b copper nitrate + ________ → iron nitrate + ________ [2 marks]

c Ca + $MgSO_4$ → ________ + ________ [2 marks]

2. A student adds grey magnesium metal to blue copper(II) sulfate solution.

Describe how the appearance of the magnesium metal would change. Explain why this change would occur.

____

____

____ [3 marks]

Extraction of metals

1. Use the reactivity series to answer these questions.

Most metals are found in the Earth as compounds. Chemical reactions are needed to extract the metals from their compounds.

a Explain why pure gold is found in the ground.

____

____ [2 marks]

b Zinc is extracted by heating its compounds with carbon. Aluminium is extracted using electrolysis.

Explain why the methods of extraction for these two metals are different.

____

____

____

____ [4 marks]

Oxidation and reduction in terms of electrons

1. **Higher Tier only**

Which statement correctly describes what happens during oxidation?

Tick **one** box.

- ☐ A metal atom gains electrons to form ions.
- ☐ A metal ion loses electrons to form atoms.
- ☐ A non-metal atom gains electrons to form ions.
- ☐ A non-metal ion loses electrons to form atoms.

[1 mark]

Remember

OILRIG:

Oxidation **I**s electron **L**oss; **R**eduction **I**s electron **G**ain

2. For each of the following half equations, write if oxidation or reduction is taking place.

a $Mg^{2+} + 2e^- \rightarrow Mg$

____________________ [1 mark]

b $2O^{2-} \rightarrow O_2 + 4e^-$

____________________ [1 mark]

c $2Cl^- - 2e^- \rightarrow Cl_2$

____________________ [1 mark]

3. A student carried out a reaction between zinc chloride and magnesium.

$ZnCl_2(aq) + Mg(s) \rightarrow MgCl_2(aq) + Zn(s)$

a Write the ionic equation for the reaction.

_______________________________ [2 marks]

b Write what was reduced and what was oxidised during the reaction.

Reduced ________________ Oxidised ________________ [2 marks]

Remember

To write an ionic equation, first write out all the atoms and ions present.

Cross out the atoms or ions on both sides that do not change during the reaction.

Then re-write the equation using the parts that remain.

Reactions of acids with metals

1. A student adds magnesium to hydrochloric acid.

a Write state symbols to complete the balanced symbol equation.

$Mg(\_\_\_\_) + 2HCl(\_\_\_\_) \rightarrow MgCl_2(aq) + H_2(\_\_\_\_)$ [1 mark]

Worked Example

Higher Tier only

b Write half equations for this reaction. [2 marks]

$Mg \rightarrow Mg^{2+} + 2e^-$ [1 mark]

$2H^+ + 2e^- \rightarrow H_2$ [1 mark]

Analysing questions

Write the symbol equation as an ionic equation. Then, split the ionic equation into two half equations and balance the charges with electrons.

c Explain why this reaction is an example of a redox reaction.

____

_______________________________ [1 mark]

A redox reaction is one where both oxidation and reduction are taking place. You can see this in the two half equations.

2. A student reacts magnesium with sulfuric acid. Name the two products formed.

a ____________________ ____________________ [2 marks]

Higher Tier only

b In this reaction hydrogen is reduced. Explain what this statement means.

____ [1 mark]

Neutralisation of acids and making salts

1. What is the formula of the salt magnesium nitrate?

Tick **one** box.

- ☐ $MgNO_3$
- ☐ $Mg(NO_3)_2$
- ☐ Mg_2NO_3
- ☐ $2MgNO_3$

[1 mark]

Remember

The second part of the salt's name comes from the acid:

- nitric acid makes nitrates; nitrate ions are NO_3^-.

2. Sodium hydroxide is an alkali. Sodium oxide is a base. Both alkalis and bases can neutralise acids.

a Describe the difference between a base and an alkali.

____ [1 mark]

b A student adds sodium oxide to hydrochloric acid. Complete the word equation for this reaction.

sodium oxide + hydrochloric acid → ____________________ + ______________ [2 marks]

3. Describe how you could find out if a powder is a metal carbonate.

____

____

____

____ [4 marks]

Making soluble salts

1. A student was asked to prepare a pure, dry sample of the salt zinc(II) sulfate by reacting a metal oxide with an acid.

Required practical

a Name the metal oxide and the acid the student should use.

metal oxide ______________________ acid ______________________ [2 marks]

b The student put the acid in a beaker and then heated it using a Bunsen burner. They added the metal oxide to the acid until the metal oxide was in excess. Describe how the student knew that the metal oxide was in excess.

____ [1 mark]

c The student now has a salt solution with an excess of metal oxide. Describe how the student should make pure, dry zinc(II) sulfate crystals. You should mention the purpose of each step.

____

____

____

____

____ [4 marks]

pH and neutralisation

1. What is the ionic equation for a neutralisation reaction?

Tick **one** box.

☐ $H_2O(l) \rightarrow 2H^+(aq) + O^{2-}(aq)$ ☐ $H_2O(l) \rightarrow 2H^-(aq) + O^{2+}(aq)$

☐ $H^+(aq) + OH^-(aq) \rightarrow H_2O(l)$ ☐ $H^-(aq) + OH^+(aq) \rightarrow H_2O(l)$ [1 mark]

2. A student added universal indicator to a range of solutions with different pH values. Complete the table.

Solution	Colour with universal indicator	pH
Neutral	Green	______
______	Orange	4
Strong alkali	______	10
Strong acid	Red	______

[4 marks]

3. A student added sulfuric acid to sodium hydroxide.

Synoptic

a Write numbers in the spaces to balance the symbol equation for the reaction.

______ $H_2SO_4(aq)$ + ______ $NaOH(aq)$

→ ______ $Na_2SO_4(aq)$ + ______ $H_2O(l)$ [2 marks]

Remember

You do not need to add any 1's when balancing an equation.

b Describe how an indicator can be used to show when all the sodium hydroxide has reacted with the sulfuric acid.

____

____

____ [3 marks]

Titrations

1. You are provided with the following equipment:

Required practical

- $25\ cm^3$ pipette and safety filler
- burette, small funnel and clamp stand
- conical flask
- white tile
- dilute hydrochloric acid
- sodium hydroxide solution
- methyl orange indicator

Describe a method that uses titration to accurately determine the volume of hydrochloric acid that will neutralise 25 cm^3 of 0.5 mol/dm^3 sodium hydroxide.

____

____

____

____

____

______________________________ [6 marks]

2. One student found that 25.0 cm^3 of 0.5 mol/dm^3 sodium hydroxide solution was neutralised by exactly 21.5 cm^3 of hydrochloric acid solution.

Maths

Higher Tier only

The balanced symbol equation for the reaction is:

$HCl(aq) + NaOH(aq) \rightarrow NaCl(aq) + H_2O(l)$

Calculate the concentration, in mol/dm^3, of the hydrochloric acid solution used.

Give your answer to 2 significant figures.

[4 marks]

Analysing questions

There are a number of steps you need to take:

1. Calculate the moles of sodium hydroxide used.

   The equation you need is:

   moles = concentration × volume

   Make sure you convert all volumes to dm^3.

2. Use the balanced equation to see the ratio of moles of alkali to moles of acid.

3. Calculate the concentration of hydrochloric acid.

   The equation you need is

   $$\text{concentration} = \frac{\text{moles}}{\text{volume}}$$

   Remember the units and remember to show all your working.

Strong and weak acids

1. Which of the following are weak acids?

Higher Tier only

Tick **two** boxes.

☐ Carbonic acid ☐ Ethanoic acid

☐ Hydrochloric acid ☐ Nitric acid [2 marks]

2. Look at the information in the table.

Concentration of hydrogen ions in mol/dm^3	pH
10^{-1}	1
10^{-2}	2
10^{-3}	3
10^{-4}	4

a Describe the relationship between the concentration of hydrogen ions and pH.

____

____

____ [1 mark]

Maths **b** Calculate the pH of a solution that contains 0.000 000 1 mol/dm^3 of hydrogen ions.

[2 marks]

Analysing questions

When asked to describe a relationship look for a pattern in the data. Here, what happens to pH when the concentration of hydrogen ions decreases?

Remember: 10^{-3} is smaller than 10^{-2}.

3. A student adds a 2 cm strip of magnesium metal to a weak acid. A reaction takes place.

a Predict how the reaction would be different when the same amount of magnesium is added to a strong acid of the same concentration.

____ [1 mark]

Command words

When you are asked to 'predict' something, you need to write what you think the outcome will be.

b Explain why the strong acid gives a different result to the weak acid.

____

____

______________________________ [3 marks]

The process of electrolysis

1. Which of the following are electrolytes?

Tick **two** boxes.

☐ Copper(II) sulfate solution ☐ Molten lead bromide

☐ Crude oil ☐ Molten sugar [2 marks]

Remember

Electrolytes are liquids that are able to conduct electricity. They contain ions that are free to move.

2. Electrolysis is used to separate an electrolyte into elements.

a Label the diagram to show the equipment needed to carry out electrolysis. [4 marks]

Higher Tier only **b** Electrons are transferred to ions at the negative electrode. Is this oxidation or reduction?

______________ [1 mark]

3. A student carries out electrolysis on copper chloride solution. Name the electrode where copper will be produced and explain why copper will be produced here.

____

____

____

______________________________ [4 marks]

Electrolysis of molten ionic compounds

1. A teacher uses the equipment below to show the electrolysis of molten zinc chloride.

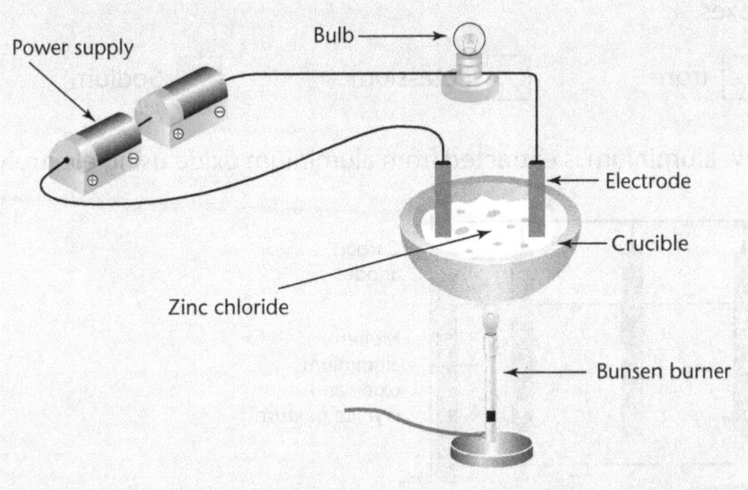

Higher Tier only

Explain what will be observed as the zinc chloride melts.
You should describe what happens to the bulb and what happens in the crucible.
Include reasons for each of your observations.

____

____

____

____

____

____ [6 marks]

2. Electrolysis is carried out on molten lead(II) bromide ($PbBr_2$).

Write the half equations for the reactions occurring at each electrode.

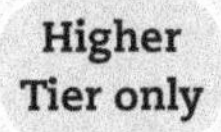

Cathode $Pb^{2+} + 2e^-$ [1 mark] $\rightarrow Pb$ [1 mark]

Anode ______________________ [4 marks]

> **Remember**
> The half equation $X^- - e^- \rightarrow X$ can also be written as $X^- \rightarrow X + e^-$

Using electrolysis to extract metals

1. Aluminium is extracted from its ores using electrolysis. Which other metals are extracted in the same way? Tick **two** boxes.

☐ Gold ☐ Iron ☐ Potassium ☐ Sodium [2 marks]

2. The diagram shows how aluminium is extracted from aluminium oxide using electrolysis.

Synoptic

a Explain why aluminium cannot be extracted by heating aluminium oxide with carbon.

____ [1 mark]

b The electrolyte is a mixture of aluminium oxide and cryolite. Explain why a mixture is used.

____

____

____ [3 marks]

c Aluminium ions and oxygen ions move to the electrodes. Predict what is discharged at each electrode. Give reasons for your answers.

Cathode ____

____

Anode ____

____ [4 marks]

Electrolysis of aqueous solutions

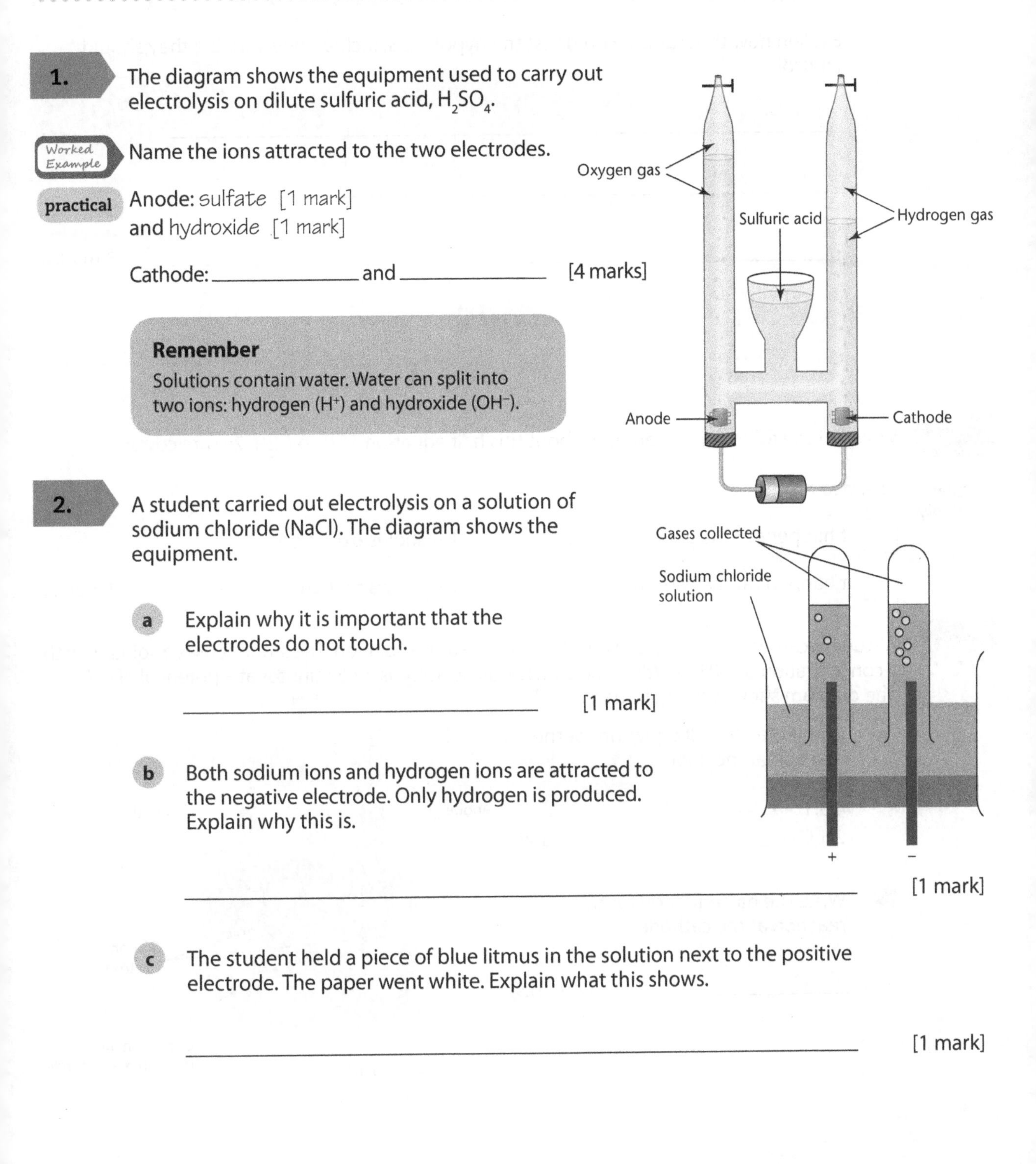

1. The diagram shows the equipment used to carry out electrolysis on dilute sulfuric acid, H_2SO_4.

Worked Example Name the ions attracted to the two electrodes.

practical Anode: sulfate [1 mark]
and hydroxide [1 mark]

Cathode: ______________ and ______________ [4 marks]

> **Remember**
> Solutions contain water. Water can split into two ions: hydrogen (H^+) and hydroxide (OH^-).

2. A student carried out electrolysis on a solution of sodium chloride (NaCl). The diagram shows the equipment.

a Explain why it is important that the electrodes do not touch.

______________________________ [1 mark]

b Both sodium ions and hydrogen ions are attracted to the negative electrode. Only hydrogen is produced. Explain why this is.

______________________________ [1 mark]

c The student held a piece of blue litmus in the solution next to the positive electrode. The paper went white. Explain what this shows.

______________________________ [1 mark]

d The student writes a hypothesis:

The rate of electrolysis will increase as the voltage increases.

Explain how the student could test the hypothesis. Include one variable they should control.

______

______

______ [3 marks]

Half equations at electrodes

1. Which of the following statements about this half equation $2Cl^- \rightarrow Cl_2 + 2e^-$ are correct?

Higher Tier only

Tick **two** boxes.

☐ It happens at the anode. ☐ It shows oxidation.

☐ It happens at the cathode. ☐ It shows reduction. [2 marks]

2. A student carries out the electrolysis of copper(II) sulfate solution. The student uses a solution with a concentration of 0.05 mol/dm^3 and carries out electrolysis for 5 minutes at a potential of 4 V. The diagram shows the equipment used.

a Complete the half equation for the reaction at the anode.

$4OH^- \rightarrow$ ______ +
$2H_2O$ + ______ [2 marks]

b Write the half equation for the reaction at the cathode.

______ [2 marks]

Synoptic **c** The student wants to investigate how the current affects the mass of copper that is deposited. Write a suitable prediction, including a scientific explanation.

____

____ [2 marks]

d Suggest how the student can check the repeatability of the results.

____

____ [2 marks]

Exothermic and endothermic reactions

1. Energy is conserved in all chemical reactions. This means that the total amount of energy in the universe is the same at the end of a reaction as it was at the start. Draw **one** line from each statement to the type of reaction. [4 marks]

Statement	Type of reaction
A reaction in which the temperature increases	Exothermic reaction
A reaction in which the temperature decreases	
Energy is transferred from the surroundings	
Energy is transferred to the surroundings	Endothermic reaction

2. A student was investigating the energy changes in a chemical reaction. Here are the results.

Maths

Temperature at start = 22 °C

Temperature at end = 5 °C

Remember

The sign is important when calculating the temperature change. If the temperature has decreased, then the change will have a negative sign.

a Calculate the temperature change.

−17°C

[1 mark]

b Identify the type of reaction taking place.

Endothermic

[1 mark]

3. A student was investigating the energy changes that take place when hydrochloric acid reacts with sodium hydroxide.

Required practical

The student placed 25 cm^3 of sodium hydroxide in a polystyrene cup and recorded the starting temperature. The student then added 5 cm^3 portions of hydrochloric acid, recorded the new temperature after each addition, and repeated until 40 cm^3 of acid had been added.

The graph shows the results.

Temperature (°C)
36 32 30 28 26 24 22 20 0
5 10 15 20 25 30 35 40
Volume of hydrochloric acid added (cm^3)

a Draw two lines of best fit on the graph. [2 marks]

b Describe how the temperature changes during the experiment.

the temp increases to a max. then starts to drop slowly. [2 marks]

c Use the graph to determine the maximum temperature reached during the experiment.

31°C [1 mark]

d Identify the type of reaction taking place. Suggest a reason for your answer.

Exothermic, the temperature of the surrandings increases. [2 marks]

e Suggest a reason why the temperature starts to decrease in the final stages of the experiment.

The reaction is finished so no more heat is added.

There is an excess of HCL, and all of the Sodium Hydroxide had reacted. [2 marks]

Reaction profiles

1. The diagrams show the reaction profiles for reactions X and Y.

X

Energy
Reactants
Products
Progress of reaction
activation energy
energy given out

Y

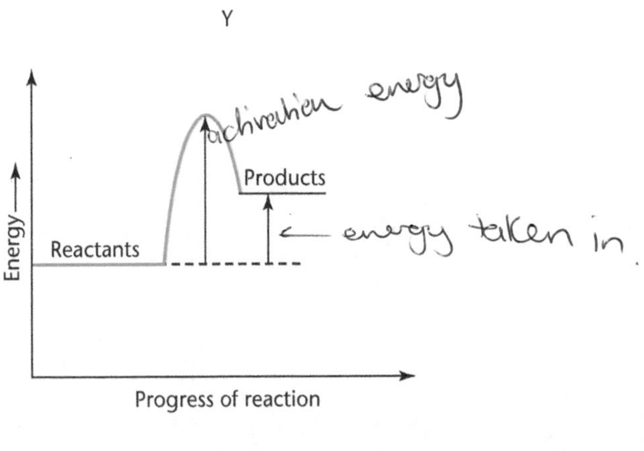

a Complete diagram X by adding labels to show the activation energy for reaction X and the amount of energy given out during the reaction. [2 marks]

b Complete diagram Y by adding labels to show the activation energy for reaction Y and the amount of energy taken in during the reaction. [2 marks]

c Explain which reaction profile is from an endothermic reaction.

Y is endothermic because the products have more energy than the reactants because energy is put INTO the system from the surroundings.

[3 marks]

2. Self-heating cans use exothermic chemical reactions to heat up their contents. The chemicals mix when a seal is broken in the can. The reaction between calcium oxide and water can be used to heat up drinks in this way.

a Complete and label the reaction profile for this reaction.

[3 marks]

b Explain what is meant by the term 'activation energy'.

The ~~minium~~ minimum energy needed to start a reaction.

[2 marks]

Energy change of reactions

1. The diagram shows a reaction profile. The reaction profile represents this reaction:

Higher Tier only

$H_2(g) + Cl_2(g) \rightarrow 2HCl(g)$

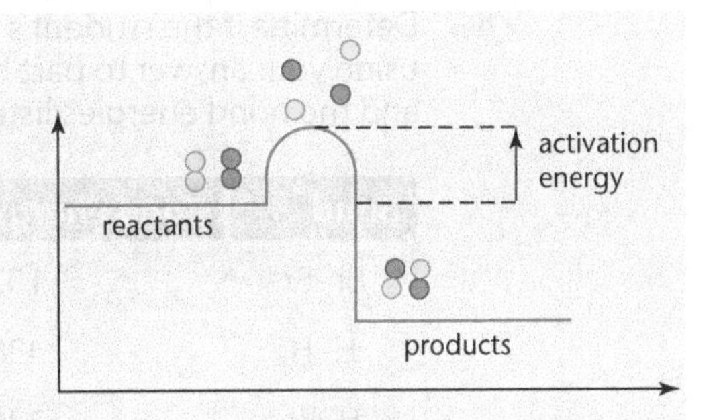

a Describe what happens during the reaction in terms of bond making and bond breaking.

____

____

____ [3 marks]

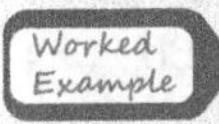

b The table lists some bond energies.

Bond	Bond energy (kJ/mol)
Cl–Cl	239
H–H	436
H–Cl	427

Calculate the overall energy change for the reaction:

$H_2(g) + Cl_2(g) \rightarrow 2HCl(g)$ [5 marks]

Work out the bond energies of the reactants from the table:

H–H = 436; Cl–Cl = 239 [1 mark]

total bond energies for reactants
= 675 kJ/mol [1 mark]

Work out the bond energies for the products:

2H–Cl = 2 × 427 [1 mark]

total bond energies for products
= 854 kJ/mol [1 mark]

Work out the difference between the bond energies of the reactants and products:

675 – 854 = –179 kJ/mol [1 mark]

Maths

If you make a mistake in the first part of the calculation, you will still be awarded marks for the rest of the calculation, provided you make no more mistakes.

Remember

The negative sign indicates that heat energy has been given out to the surroundings, therefore the reaction is exothermic.

c The reactivity of the halogens decreases as you go down Group 7.

A student predicts that less energy is given out when H_2 reacts with Br_2 to produce HBr when compared to the reaction between H_2 and Cl_2.

Complete the equation for the reaction.

$H_2(g) + Br_2(l) \rightarrow$ ________(____) [1 mark]

d Determine if the student's prediction is correct by using your answer to part b, your equation in part c and the bond energies listed in the table.

Bond	Bond energy (kJ/mol)
Br–Br	193
H–H	436
H–Br	336

Command words

'Determine' means to use the given data or information to obtain an answer.

Follow the method of the worked example in part b. Then compare the two values.

Maths

Remember that a smaller negative number is larger than a larger negative number. For example, –6 is larger than –100.

____

____

____

____

____

____

____

______________________ [6 marks]

Cells and batteries

1. Use words from the box to complete the sentence.

batteries	cells	current	parallel	series	voltage

______________ consist of two or more ______________ connected together in ______________ to provide a greater ______________. [4 marks]

2. The diagram shows an electrochemical cell.

a Describe what will happen when one electrode is lifted out of the electrolyte.

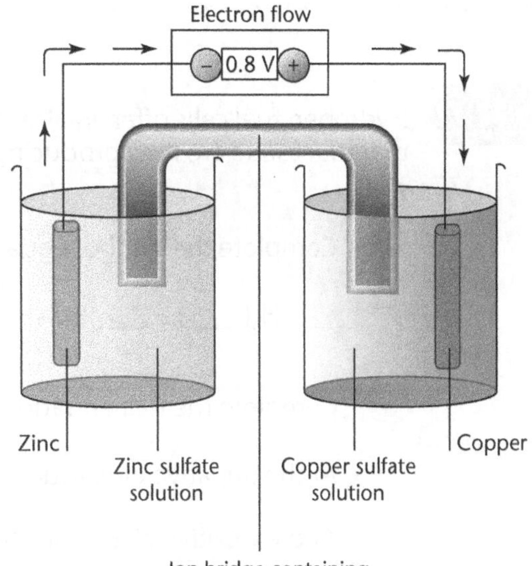

____

____

____

____

____

____________________________ [1 mark]

> **Command words**
> 'Describe' means 'tell what you see'. You do not need to give a reason for your observations.

Higher Tier only

b The following reactions are taking place in the cell. Balance the half equations.

$Zn \rightarrow Zn^{2+} +$ ____ e^- $\quad\quad$ $Cu^{2+} +$ ____ $e^- \rightarrow Cu$ [2 marks]

c Explain why the cell contains an ion bridge.

____ [1 mark]

> **Command words**
> 'Explain' means to make clear or state the reasons for something happening.

d Predict what will happen to the voltage of the cell if the zinc electrode is replaced by a magnesium electrode in a solution of magnesium sulfate.

____ [1 mark]

> You will need to use the reactivity series when making your prediction.

> **Command words**
> 'Predict' means to give a plausible suggestion based on your chemical knowledge.

Fuel cells

1. Hydrogen fuel cells offer an alternative to rechargeable cells and batteries. Some car manufacturers are now producing cars that are powered by hydrogen fuel cells.

a Complete the symbol equation, including state symbols.

____H_2(____) + ____(g) → ____H_2O(____) [3 marks]

b Complete the half equation for each electrode in the fuel cell.

At the negative electrode: ____ → ____H^+ + ____e^-

At the positive electrode: $4H^+ + O_2$ + ____e^- → ____H_2 [2 marks]

Remember
The number of electrons in the two balanced half equations cancel each other out.

c Describe **two** advantages of using hydrogen fuel cells to power cars.

____

____ [2 marks]

d Describe **two** disadvantages of using hydrogen fuel cells to power cars.

____

____ [2 marks]

Measuring rates of reaction

1. Which of these reactions is the quickest?

Tick **one** box.

☐ A nail rusting ☐ Concrete setting ☐ Paint drying

☐ Reaction between marble chips and hydrochloric acid [1 mark]

2. Hydrogen gas is produced during the reaction between hydrochloric acid and magnesium.

a Write the word equation for this reaction.

______________________________ [1 mark]

A student wants to measure the rate of this reaction. The diagram shows how the student set up the apparatus.

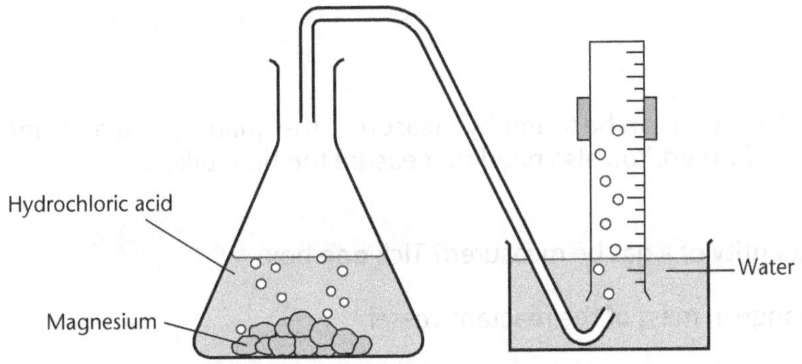

Hydrochloric acid

Magnesium

Water

b Identify **two** errors in the way the apparatus is set up.

____

______________________________ [2 marks]

c Describe what would happen if the student used the apparatus shown in the diagram above.

____

______________________________ [2 marks]

The student corrected the errors and then carried out the experiment. The student collected some data every 10 seconds and the results are plotted in the graph.

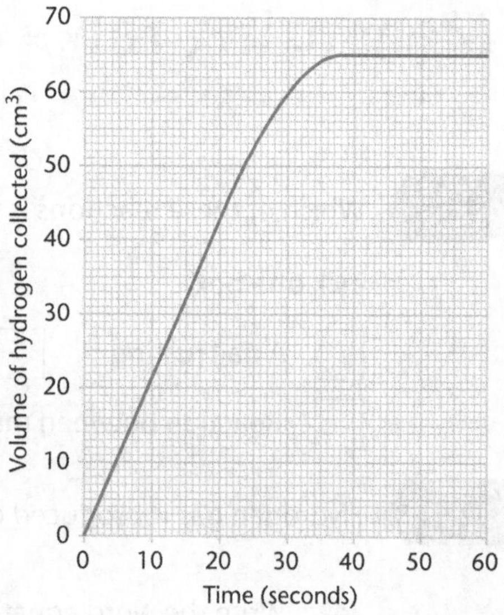

d When was the reaction at its fastest?

______________________ [1 mark]

e How much hydrogen was produced during the reaction?

______________________ [1 mark]

Volume of hydrogen collected (cm^3)

Time (seconds)

f State the time taken for the reaction to complete.

______________________ [1 mark]

Calculating rates of reaction

1. The rate of a chemical reaction can be found by measuring the quantity of a reactant used or the quantity of product formed. You also need to measure the time taken.

a How can the quantity of a gas be measured? Tick **one** box.

- ☐ The change in mass of the reactant vessel
- ☐ The change in temperature during the reaction
- ☐ The change in colour of the solution
- ☐ The change in mass of the catalyst

[1 mark]

b Write down the equation that links the mean rate of reaction with the time taken and the quantity of gas produced.

______________________ [1 mark]

If you read the question carefully you will find that much of the answer is hidden there. Highlight the words 'rate', 'produced' and 'time taken' and then re-write the sentence as an equation. You will not be given this equation in an exam. If the rate is given in g/s then you need to divide a mass (in g) with the time (in s).

Maths **c** Some nitric acid was added to a sample of magnesium metal. 20 cm^3 of gas was collected in the first 40 seconds. Calculate the mean rate of reaction and give the correct units.

[2 marks]

2. The graph shows the total volume of hydrogen produced when zinc reacts with excess hydrochloric acid.

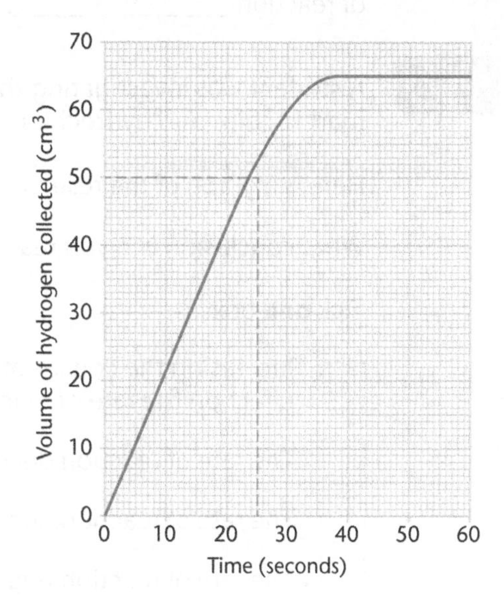

a Write on the graph where the reaction is:

1. fastest
2. slowing down
3. stopped. [3 marks]

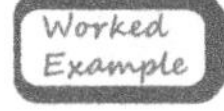

b Calculate the rate of reaction from the gradient of the graph. Give the units.

Choose a straight part of the graph and draw two construction lines (in red).

Measure the value of x and y using the scales

$x = 25$, $y = 50$

slope of the gradient = $\frac{y}{x}$, so we divide y by x

$\frac{50}{25} = 2$

Determine the units using the units given on the axes; x is in seconds (s); y is in cubic centimetres (cm^3) so the units are cm^3/s. Therefore, rate = 2 cm^3/s.

Command words

'Calculate' means to use the numbers/data from the question to work out the answer.

[3 marks]

Effect of concentration and pressure

1. Use words from the box to complete the sentences.

activation	atoms	chemical	concentration	decreases	increases
limited	particles	physical	sufficient	temperature	

Collison theory explains that ______________ reactions can occur only when reacting ______________ hit or collide with each other with ______________ energy. The minimum amount of energy is known as the ______________ energy.

As the ______________ increases, the number of particles in the same volume increases so does the number of successful collisions. Therefore, the rate of reaction ______________. [6 marks]

2. A student was investigating the reaction between marble chips and hydrochloric acid.

Here is a graph of the results.

What conclusion can you reach from this experiment?

Tick **one** box.

☐ Increasing the concertation of the acid has no effect on the rate of reaction.

☐ The rate of reaction decreases as the concentration of the acid increases.

☐ The rate of reaction depends on the change in mass.

☐ The rate of reaction depends on the concentration of the acid. [1 mark]

3. A student is investigating the rate of reaction between sodium thiosulfate and hydrochloric acid. The equation for the reaction is

$$Na_2S_2O_3(aq) + 2HCl(aq) \rightarrow S(s) + 2NaCl(aq) + SO_2(g) + H_2O(l)$$

a Describe how the student could measure the rate of reaction.

____

____

____ [3 marks]

Worked Example

b Predict how the rate of the reaction will change if the student doubles the concentration of the acid. Give reason for your answer.

The rate of reaction will double. [1 mark] There will be twice as many hydrochloric acid particles in the beaker [1 mark] which means it will be likely that there will be twice as many successful collisions with the sodium thiosulfate particles per second. [1 mark] [3 marks]

4. The diagram shows a gas syringe containing particles of two different of gases.

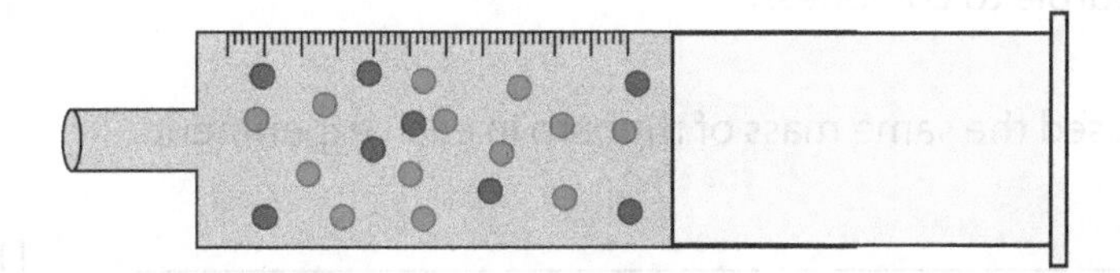

Explain why the rate of reaction will decrease when the plunger is pulled out.

____

____

____ [3 marks]

Effect of surface area

1. The diagram shows some reacting particles of calcium carbonate and hydrochloric acid.

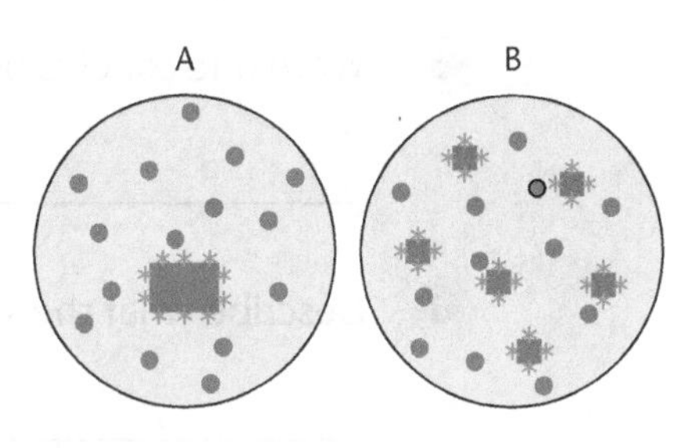

a Write which diagram shows the particles with the smallest total surface area. Give a reason for your answer.

____ [2 marks]

b Predict which reaction will take place at the fastest rate. Give a reason for your answer.

____

____ [2 marks]

2. A student investigated the rate of reaction between oxalic acid, which is found in rhubarb, and acidified potassium permanganate.

The student poured 30 cm^3 of acidified potassium permanganate into three conical flasks and then added the same mass of rhubarb.

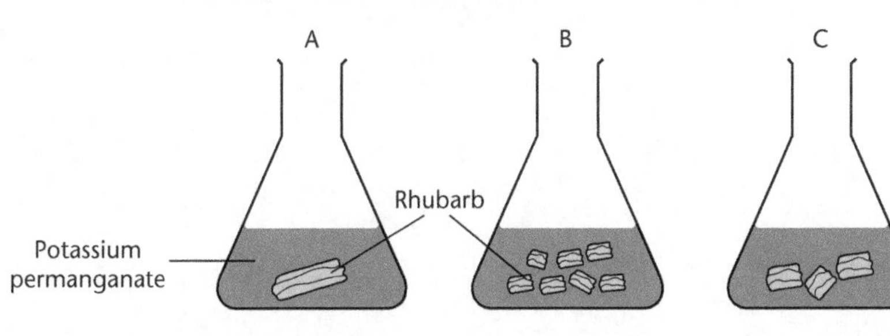

The student gently stirred each beaker and timed how long it took for the potassium permanganate to turn from purple to colourless.

a Write why the student used the same mass of rhubarb in each experiment.

______________________________ [1 mark]

b Give the independent variable in this experiment.

______________________________ [1 mark]

The table shows the results.

Conical flask	Time taken for the colour change (s)
A	100
B	20
C	64

c Write one conclusion the student can make from these results.

______________________________ [1 mark]

d Describe what the student could do to increase confidence in the conclusion.

____

______________________________ [2 marks]

Effect of temperature

1. A precipitate is formed when sodium thiosulfate solution is mixed with hydrochloric acid.

a A student investigated the effect of temperature on the rate of reaction between sodium thiosulfate and hydrochloric acid. They mixed together the reactants and placed the conical flask on a black cross. They measured the time take for the black cross to disappear.

The table shows the results. Plot a graph of the results.

Temperature (°C)	Time (s)
20	200
30	175
40	146
50	111
60	80

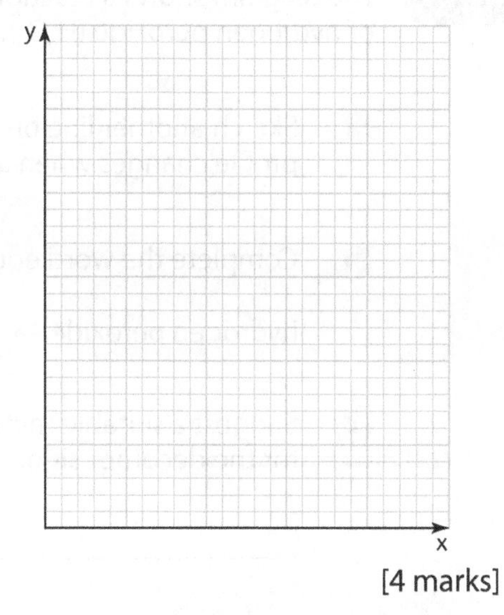

[4 marks]

b Write a conclusion that can be drawn from the graph. Explain your answer.

You need to use the particle theory here.

____

____

____

______________________________ [4 marks]

c Suggest what the student should do to improve the results.

____

______________________________ [2 marks]

Effect of a catalyst

1. Which of the following statements are true?

Tick **two** boxes.

- [ ] Catalysts are specific to one single reaction.
- [ ] Catalysts are substances that change the speed of a chemical reaction.
- [ ] Catalysts are used up in the reaction.
- [ ] Catalysts are usually used in large amounts.

[2 marks]

2. The diagram shows a reaction profile for the decomposition of hydrogen peroxide (H_2O_2).

Command words
'Sketch' means to draw an approximate line.

a Sketch another line on the graph to show how the profile changes when a catalyst is added. [1 mark]

Energy
Activation energy
Reactants
Products
Progress of reaction

b Complete the word equation for the reaction.

hydrogen peroxide → ________ + ________ [1 mark]

c Suggest a suitable method to find a suitable catalyst for the decomposition of hydrogen peroxide.

____

____

______________________________ [3 marks]

Some results from this experiment are shown in the table.

Experiment	Substance added	Time to collect gas (s)
1	No substance	121
2	Liver	75
3	Manganese oxide	12
4	Iron oxide	24

Command words
'Suggest' means to apply your knowledge and understanding of science to a new situation or context.

d Explain why no substance was added in experiment 1.

______________________________ [1 mark]

e Suggest a reason why liver was used in experiment 2.

______________________________ [1 mark]

f Identify the best catalyst for this reaction. Explain your answer.

Command words
'Identify' means to name or otherwise characterise.

____

______________________________ [2 marks]

Reversible reactions and energy changes

1. In some chemical reactions, the products of the reaction can react to produce the original reactants. Name this type of reaction. Tick **one** box.

☐ Irreversible ☐ Redox ☐ Reversible ☐ Thermal decomposition [1 mark]

2. Look at this equation:

$$\text{hydrated copper sulfate (blue)} \underset{\text{exothermic}}{\overset{\text{endothermic}}{\rightleftharpoons}} \text{anhydrous copper sulfate (white)} + \text{water}$$

a Give the meaning of the $\rightleftharpoons$ symbol.

______________________________ [1 mark]

b Describe the conditions required for the forward reaction to take place. Suggest a reason for your answer.

____

____

______________________________ [3 marks]

c Describe the conditions required for the reverse reaction to take place.

______________________________ [1 mark]

2. Literacy

A sample of white ammonium chloride was placed in a boiling tube. On heating, the amount of solid in the boiling tube reduced. However, a white solid was observed to form around the top of the boiling tube. Explain these observations.

____

____

____

______________________________ [4 marks]

Equilibrium and Le Chatelier's Principle

1. Complete the sentence.

When a reversible reaction occurs in a closed system, equilibrium is reached when

______________________________ [1 mark]

2. The reaction $A + B \rightleftharpoons C + D$ is at equilibrium. The concentrations of A and B are less than the concentrations of C and D. Predict where the position of equilibrium lies.

______________________________ [1 mark]

3. In a sealed bottle of fizzy pop, the following equilibrium is in place:

$CO_2(g) \rightleftharpoons CO_2(aq)$

Suggest why a fizzing noise is heard when the bottle lid is unscrewed.

> Look at the equation. Equilibrium occurs when there is a reversible reaction *in a closed system*. If these conditions are not met, then the reaction stops being in equilibrium.

____

____

______________________________ [3 marks]

4. Higher Tier only

Write Le Chatelier's Principle as applied to a reaction at equilibrium.

____

______________________________ [1 mark]

5. Higher Tier only

Dinitrogen tetroxide is in equilibrium with nitrogen dioxide. The equation for the reaction is:

$N_2O_4(g) \rightleftharpoons 2NO_2(g)$

(colourless) (brown)

The forward reaction is endothermic.

> Look at the equation to find out the colours of each gas. When predicting an observation remember to write down the starting colour as well as the final colour.

a Predict what you will observe when the pressure is increased.

______________________________ [1 mark]

b Predict what you will observe when the temperature is increased.

______________________________ [1 mark]

You need to learn Le Chatelier's Principle and be able to apply it to different situations.

Changing the position of equilibrium

1. The Haber process produces ammonia from nitrogen and hydrogen. The reaction is exothermic.

Higher Tier only

The Haber process is also covered in Section 10.

a Complete the equation by adding the 'balancing' numbers.

$N_2(g)$ + ____ $H_2(g) \rightleftharpoons$ ____ $NH_3(g)$ [2 marks]

In the Haber process reactor, the nitrogen and hydrogen are heated at high pressures and passed over a catalyst.

b Changing the reaction conditions will alter the position of the equilibrium.

Describe how you could alter the reaction conditions to **increase** the yield at equilibrium.

First work out which way the equilibrium needs to be moved to **increase** the yield.

____

______________________________ [2 marks]

Worked Example

c Explain why ammonia is liquefied and removed as soon as it is formed. [2 marks]

Ammonia is removed so more ammonia will be formed. [1 mark]
In an attempt to restore the equilibrium more hydrogen and nitrogen will react. [1 mark]

Worked Example

d To obtain the most economic yield for the Haber process, a compromise must be made between the pressures and temperature used. Explain why this is. [4 marks]

The forward reaction favours low temperatures [1 mark] but if the temperature is too low the rate of reaction will be very slow [1 mark] and the reaction becomes economically unviable.

The forward reaction favours high pressures [1 mark] but these require very expensive equipment. [1 mark] So in practice the industry tries to get a good balance between rate, cost and yield.

2. Ethanol can be manufactured by reacting ethene with steam. The reaction is reversible and the formation of the ethanol is exothermic.

> To find the formula of ethanol you need to look in Section 7, 'The structure and formulae of alcohols'.

a Complete the equation:

$C_2H_4(g) + H_2O(g)$ ____________________ [3 marks]

b Predict what happens to the position of equilibrium when the temperature is increased. Give a reason for your answer.

____

____ [2 marks]

c Predict what happens to the position of equilibrium when the pressure is increased. Give a reason for your answer.

> Count the number of molecules that are gases on each side of the reaction.

____

____ [2 marks]

d During the reaction the vapours are passed over a phosphoric acid catalyst. Describe the effect of the catalyst on the position of equilibrium.

____ [1 mark]

e Explain why this reaction is carried out at a temperature of 300 °C and a pressure of 60–70 atmospheres.

____

____

____

____ [4 marks]

Structure and formulae of alkanes

1. Which of these molecules is an alkane? Tick **one** box.

☐ C_2H_2 ☐ C_3H_8 ☐ C_4H_{12} ☐ C_5H_8 [1 mark]

2. Molecules can be represented in different ways.

Look at this image of an alkane.

Write down the formula of the alkane. ________ [1 mark]

3. Draw **one** line from each alkane structure to the alkane name. [2 marks]

Structure

```
   H  H  H  H
   |  |  |  |
H—C—C—C—C—H
   |  |  |  |
   H  H  H  H
```

```
   H
   |
H—C—H
   |
   H
```

Name

Methane	Ethane	Propane	Butane

4. The general formula for the alkanes is C_nH_{2n+2} where *n* is the number of carbon atoms.

Maths

a Write down the formulae for an alkane containing six carbon atoms.

________ [1 mark]

b Write down the formulae for an alkane containing 25 carbon atoms.

You need to substitute the numbers into the formula. 2*n* + 2 means two times the number of carbon atoms in the molecule plus 2.

________ [1 mark]

Fractional distillation and petrochemicals

1. Crude oil is a fossil fuel. To make crude oil more useful it is separated into fractions.
Use words from the box to complete the sentences.

boiling	compound	condense	distillation	element	evaporated
filtration	freezing	melting	mixture	simple	

Crude oil is a ______________________________ of different hydrocarbons. It is

separated by the process of fractional ______________________________. The oil is

______________________________ and allowed to cool and condense. The different

fractions have different ______________________________ points and so will condense at

different temperatures. [4 marks]

2. Crude oil is separated into its components in a fractionating column. Explain why fuel oils are collected near the bottom of the column whereas paraffin is collected near the top of the column.

> To get full marks you will need to compare the two types of molecules with each other, making three points. A good place to start is to think about the relative size of the molecules.

____

____

____

______________________________ [3 marks]

3. The table shows the percentage of the fractions in crude oil from two different oil wells.

Maths **a** A barrel of crude oil from well A has a mass of 150 kg whereas a barrel of oil from well B has a mass of 142 kg. Calculate whether barrels from well A or well B contain the most paraffin.

[2 marks]

Fraction	% of fractions in crude oil	
	Well A	Well B
LPG	6	2
Petrol	10	11
Paraffin	15	20
Diesel	7	5
Heating oil	6	4
Fuel oil	14	12
Bitumen	42	46

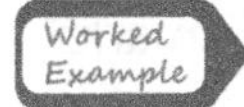

b The petrochemical industry has a high demand for fractions such as LPG and petrol. There is much less demand for fuel oil and bitumen.

Use the data from the table to explain why this difference in demand is a potential problem. [4 marks]

Fuel oil and bitumen make up 56% of the crude oil from well A and 58% of the crude oil from well B. [1 mark]

LPG and petrol only make up 16% of the crude oil for well A and 13% of the crude oil from well B. [1 mark]

So the supply of LPG and petrol is very low compared to the demand, which means the supply to the petrochemical industry could run out [1 mark]. The industry could be left with lots of fuel oil and bitumen that thy cannot use [1 mark] since the supply of this is much greater than the demand.

> **Command words**
> 'Use' means that your answer must be based on the information given in the question.

Properties of hydrocarbons

1. Some properties of hydrocarbons depend on the size of their molecules. Which of the following statements are true? Tick **two** boxes.

☐ A hydrocarbon with 20 atoms is more flammable that one with nine carbon atoms.

☐ As the number of carbon atoms in a hydrocarbon molecule increases, its melting point decreases.

☐ Large hydrocarbon molecules are more viscous than small hydrocarbon molecules.

☐ Methane is easier to ignite than butane. [2 marks]

2. The table shows the boiling points of the first six alkanes.

Alkane	Number of carbon atoms	Boiling point (°C)
Methane	1	−161
Ethane	2	−88
Propane	3	−42
Butane	4	−0.5
Pentane	5	
Hexane	6	68

Worked Example **a** Plot the data on the axes below. [3 marks]

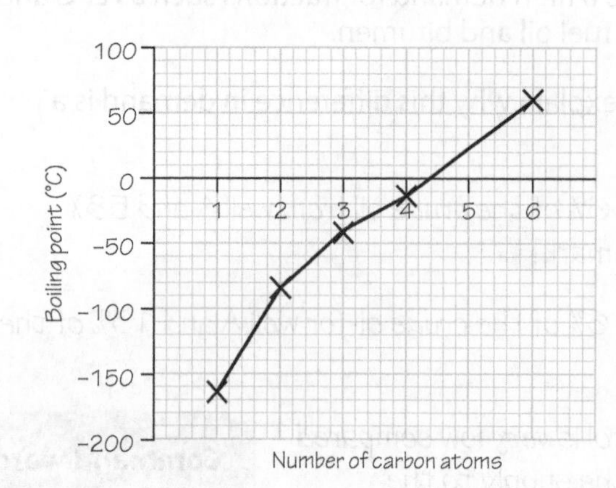

b Predict the boiling point of pentane.

______________ [1 mark]

Draw a line up from '5' on the horizontal axis to the line of best fit. Then draw a line across to the vertical axis. What are the units?

c Describe the general pattern shown by the graph.

______________ [1 mark]

d Explain the general trend in the boiling points of the alkanes.

____

______________ [2 marks]

Use the particle theory of matter. Don't forget the weak forces that exist *between* particles.

Combustion of fuels

1. Incomplete combustion occurs when hydrocarbon fuels burn in a limited oxygen supply. Which of these are products of incomplete combustion? Tick **two** boxes.

☐ Carbon ☐ Carbon dioxide ☐ Carbon monoxide [1 mark]

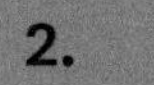

2. Write a balanced equation for the complete combustion of butane, C_4H_{10}.

Higher Tier only

First write down the formulae you need in the symbol equation: [3 marks]

$C_4H_{10} + O_2 \rightarrow CO_2 + H_2O$

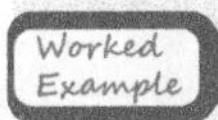

Next count up the atoms on each side of the equation:

LHS $4 \times C, 10 \times H, 2 \times O$ RHS $1 \times C, 3 \times O, 2 \times H$

Now start trying to balance the number of atoms. There are 4 carbons and 10 hydrogens on the left. Balance these on the right.

$C_4H_{10} + O_2 \rightarrow \mathbf{4}CO_2 + \mathbf{5}H_2O$

Recount the number of atoms on each side:

LHS $4 \times C, 10 \times H, 2 \times O$ RHS $4 \times C, 13 \times O, 10 \times H$

The C and H are correct; now balance the oxygens. Count them up on the right and work out how many are needed on the left.

$C_4H_{10} + \mathbf{6.5}O_2 \rightarrow 4CO_2 + 5H_2O$

Finally, multiply by 2 to obtain whole numbers:

$2C_4H_{10} + 13O_2 \rightarrow 8CO_2 + 10H_2O$

This final equation would obtain full marks:

- 1 mark for writing down the formulae correctly
- 1 mark for balancing the reactants
- 1 mark for balancing the products.

Cracking and alkenes

1. Draw **one** line from each word to its meaning.

Key word	Meaning
Alkene	Process of breaking down larger hydrocarbons into more useful smaller hydrocarbons
	A chemical that speeds up a reaction but is not changed by it or used up
Cracking	A hydrocarbon that contains only single bonds
	A hydrocarbon that contains a double bond

[2 marks]

2. The laboratory apparatus used for cracking $C_{16}H_{34}$ was set up as below.

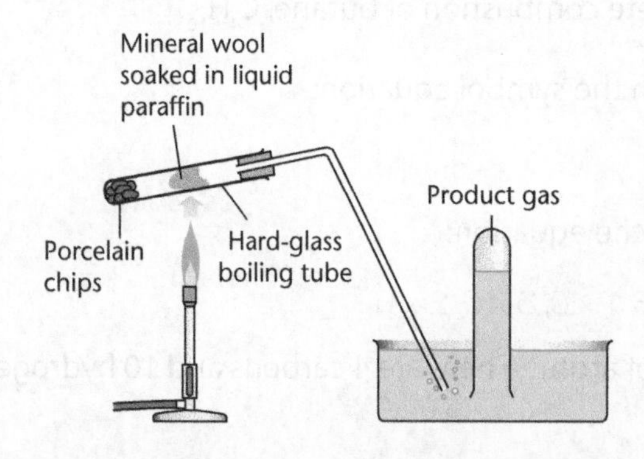

a Identify **two** mistakes in the diagram.

____

____ [2 marks]

b Complete the equation.

This equation needs to be balanced.

$C_{16}H_{34} \rightarrow C_6H_{14} + C_4H_8 +$ ____ C_2H_4 [1 mark]

c Identify the products in the equation in part b that are alkenes.

____ [1 mark]

d Describe a chemical test that will show if the cracking reaction has been successful.

____

____

____ [3 marks]

Structure and formulae of alkenes

1. Complete the table. The first row has been done for you.

Worked Example

Name	Number of carbon atoms	Formula	Displayed formula
Ethene	2	C_2H_4	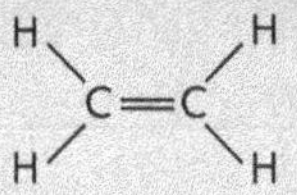 $H_2C=CH_2$
Propene	3		
Butene		C_4H_8	
Pentene			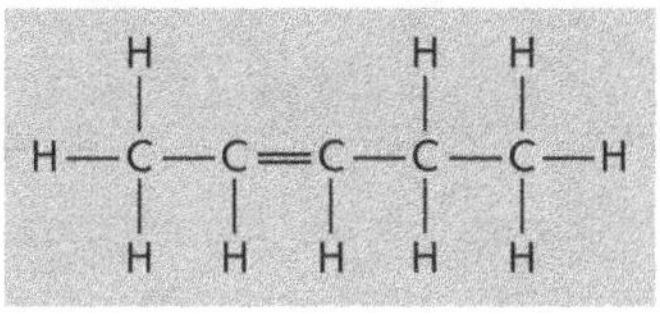 $H_3C-CH=CH-CH_2-CH_3$

[3 marks]

2. The general formula for the alkenes is C_nH_{2n} where n is the number of carbon atoms.

a Write down the formula for an alkene containing six carbon atoms.

____

[1 mark]

b Write down the formula for an alkene containing eight carbon atoms.

____

[1 mark]

3. Explain, giving a reaction as an example, the difference in reactivity between propane and propene.

Literacy

What is an important difference between the two compounds?

____

____

____

____

[3 marks]

Reactions of alkenes

1. Use words from the box to complete the sentences.

addition	double	elimination	less	more
saturated	single	unsaturated		

Alkenes are ____________ reactive than alkanes as they have a reactive ____________ bond within their carbon/hydrogen skeleton. Alkenes react by the ____________ of molecules across this double bond, so the carbon–carbon bond becomes a single bond. The unsaturated molecules become ____________. [4 marks]

2. **a** Complete the equation for the reaction between ethene and hydrogen gas.

$$H_2C{=}CH_2 + H{-}H \longrightarrow$$

[1 mark]

b Name the product that is formed.

____________ [1 mark]

c Explain why this reaction is called an addition reaction.

____________ [1 mark]

3. When propene reacts with chlorine gas, 1,2-dichloropropane is produced. Which structure shows the product? Tick **one** box.

☐ $CH_2Cl{-}CH_2{-}CH_2Cl$ (H–C(Cl)(H)–C(H)(H)–C(Cl)(H)–H)

☐ $CHCl_2{-}CH_2{-}CH_3$... as drawn: H–C(Cl)(H)–C(Cl)(H)–C(H)(H)–H

☐ H–C(Cl)(H)–C(H)(H)–C(H)(H)–H

☐ Cl–C(Cl)(H)–C(H)(H)–C(H)(H)–H

[1 mark]

4. A student was investigating the combustion of some hydrocarbons. The results are shown in the table.

Sample	CH_4	C_2H_4	C_2H_6	C_3H_6	C_3H_8
Result	Clean flame	Smoky flame	Clean flame	Smoky flame	Clean flame

a Give a conclusion that the student can draw from the results.

______________________________ [1 mark]

b Suggest a reason for your answer to part a.

______________________________ [1 mark]

c Explain which fuel you would choose to use for a camping stove.

____

______________________________ [2 marks]

Structure and formulae of alcohols

1. The diagram shows a 3D image of an alcohol.

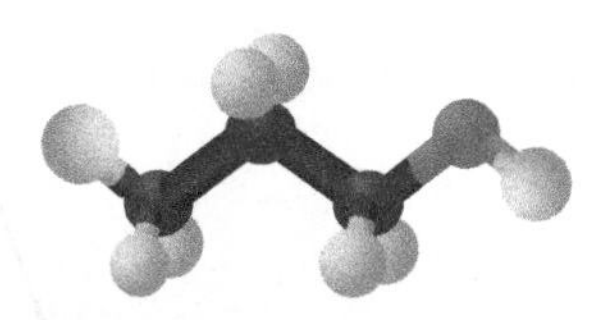

a Draw the displayed formula of this alcohol.

[1 mark]

b Give the name of this alcohol.

______________ [1 mark]

Count the number of C atoms to work out what the first part of the name is. There are just carbon–carbon single bonds, so the middle of the name is 'an'. The ending for an alcohol is 'ol'.

2. Compare the chemical structure of an alkane with the chemical structure of an alcohol.

Literacy

When comparing the structures of alkanes and alcohols, draw out displayed formulae of two specific examples so you can compare the types of bonds and the number and type of atoms present.

____

____

____

____ [4 marks]

3. Ethanol is produced when sugar solutions are fermented using yeast.
A group of students wanted to make some ethanol in the lab using the equipment shown in the diagram.

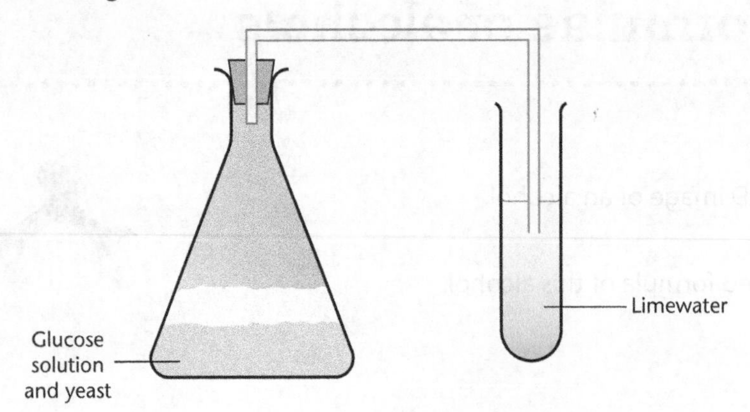

a Describe what the students will observe during the reaction.

In the conical flask? _______________________________

In the boiling tube? _______________________________ [2 marks]

b Name the gas produced during the reaction. _______________ [1 mark]

c Name the biological catalyst used in this reaction. _______________ [1 mark]

d Explain why the reaction was carried out between 25 °C and 40 °C.

____

____ [2 marks]

e Write a word equation for the fermentation reaction.

glucose → ________________ + ________________ ________________ [1 mark]

Uses of alcohols

1. Why is ethanol is widely used as a solvent?

Tick **one** box.

- [ ] It is toxic.
- [ ] Many organic substances are soluble in it.
- [ ] It is very expensive.
- [ ] Most organic substances are not soluble in it.

[1 mark]

2. Methanol is used as an industrial feedstock.

a Give the meaning of the word 'feedstock'.

____ [1 mark]

b Which of the following products does **not** use methanol as a feedstock?

Tick **one** box.

- [ ] Adhesives
- [ ] Alcoholic drinks
- [ ] Esters
- [ ] Food flavouring

[1 mark]

3. Ethanol can be used as a fuel. Ethanol burns in air to produce carbon dioxide and water.

a Write down the word equation for this reaction.

____ [1 mark]

b Complete the balancing in the symbol equation for this reaction.

$C_2H_5OH +$ ____$O_2 \rightarrow$ ____$CO_2 + 3H_2O$ [2 marks]

> You will need to make sure there are the same number of atoms on the two sides of the equation. Remember that the number in front of the molecule multiplies *all* the atoms in that molecule. $3H_2O$ means $3 \times 2 = 6$ H atoms and $3 \times 1 = 3$ O atoms.

4. Esters are sweet smelling organic compounds that are used as perfumes and food flavouring.

Literacy

Describe, giving an example, how you would make an ester.

____

____

____ [3 marks]

Carboxylic acids

1. Draw **one** line from each formula to the name of the compound.

Formula	Name
CH_3COOH	Methanoic acid
	Ethanoic acid
C_2H_5COOH	Propanoic acid
	Butanoic acid

[2 marks]

2. When ethanoic acid and potassium carbonate are mixed together a gas is given off. Complete the word equation for this reaction.

ethanoic acid + ______________________

$\rightarrow$ potassium ethanoate + ____________ + ____________ ____________ [3 marks]

3. Explain why ethanoic acid is a weak acid but hydrochloric acid is a strong acid.

Higher Tier only

Literacy

____

____

____

____ [4 marks]

4. Compare the structure and properties of carboxylic acids with those of alcohols.

Literacy

____

____

____

____

____

____

____

____

____

____ [6 marks]

When comparing the structure of carboxylic acids and alcohols, draw out displayed formula of specific examples so you can compare the bonds and functional groups. Remember that a carboxylic acid is a weak acid.

Addition polymerisation

1. Use words from the box to complete the sentence.

atoms	condensation	monomers	neutralisation
particles	polymerisation	polymers	

In addition, ______________ reactions, many small molecules called ______________ join together to form very large molecules called ______________. [3 marks]

2. The diagram shows part of a polymer molecule.

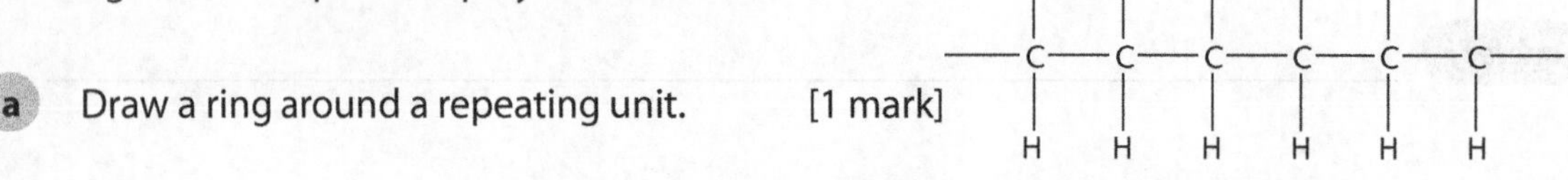

a Draw a ring around a repeating unit. [1 mark]

b Draw the displayed formula of the monomer from which this polymer is made.

[1 mark]

3. Ethene is used in the manufacture of the plastic (poly)ethene. The diagram shows what happens in the reaction.

$$n\ H_2C{=}CH_2 \longrightarrow \left[-CH_2-CH_2-\right]_n$$

a Explain what the '*n*' represents in the equation.

____ [1 mark]

b Explain why a bracket has been drawn around the product.

____ [1 mark]

Worked Example

c Explain why the reaction that produces poly(ethene) is an example of addition polymerisation. [1 mark]

Many small ethene molecules add together to form a very large molecule. [1 mark]

d Describe how the ethene molecules join together to form (poly)ethene.

____

____

____ [3 marks]

e Write down the name of the polymer that would be made if ethene was replaced by butene.

____ [1 mark]

Condensation polymerisation

1. Some polymers, such as polyester and nylon, are not formed by addition polymerisation.

Higher Tier only

a Name the type of polymerisation reaction used to form polyester and nylon.

____ [1 mark]

b Identify what is produced alongside the polymer during this type of polymerisation.

____ [1 mark]

c Which of these molecules could be used in a condensation polymerisation reaction?

Tick **one** box.

☐ CH_3CH_2OH ☐ $HO—CH_2CH_2—OH$

☐ $CH_3CH_2—CH_2$ ☐ CH_3COOH [1 mark]

Worked Example

d Explain why hexanedioic acid can take part in condensation polymerisation but ethanoic acid cannot. [2 marks]

For condensation polymerisation to occur the monomers must contain two functional groups. [1 mark] Ethanoic acid has only one acid group per molecule, but hexanedioic acid has two. [1 mark]

2. The equation shows the production of a polyester by condensation polymerisation.

$$n\text{HO}-\bigcirc-\text{OH} + n\,\text{HOOC}-\bigcirc-\text{COOH} \rightarrow \left[\bigcirc-\text{OOC}-\bigcirc-\text{COO}\right]_n + 2n\text{H}_2\text{O}$$

a Circle the functional groups in each monomer. [2 marks]

Each monomer has a two functional groups.

b Highlight the repeating unit. [1 mark]

c Explain why the reaction is called a condensation reaction.

[1 mark]

3. Poly(ethene) is made by addition polymerisation. Nylon is made by condensation polymerisation. Compare the two different polymerisation processes.

Literacy

[5 marks]

Amino acids

1. Alanine is an amino acid.

Higher Tier only

Name the type of polymerisation reaction that take place when a polypeptide is formed.

[1 mark]

2. The displayed formula for glycine in shown here.

$$H_2N-CH_2-C(=O)-O-H$$

a Draw a ring around the two functional groups in a glycine molecule. [2 marks]

b Name the small molecule that is formed during the polymerisation of glycine.

____ [1 mark]

c Write down the formula of the repeating unit of poly(glycine).

[1 mark]

3. Explain how a protein is formed from amino acids.

____

____

____

____

____

____ [6 marks]

DNA and other naturally occurring polymers

1. Which compound is the monomer that produces the natural polymer starch? Tick **one** box.

☐ Amino acid ☐ Cellulose

☐ Glucose ☐ Polysaccharide [1 mark]

2. The diagram shows a simplified structure for glucose. Glucose is a monosaccharide. A polysaccharide is formed when glucose undergoes condensation polymerisation.

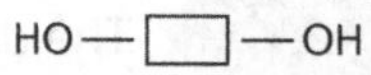

HO—☐—OH

a Draw a section of the polysaccharide chain in the space below.

[2 marks]

b Name the small molecule that is also formed during the reaction.

______________________________ [1 mark]

3. The diagram shows part of a naturally occurring polymer.

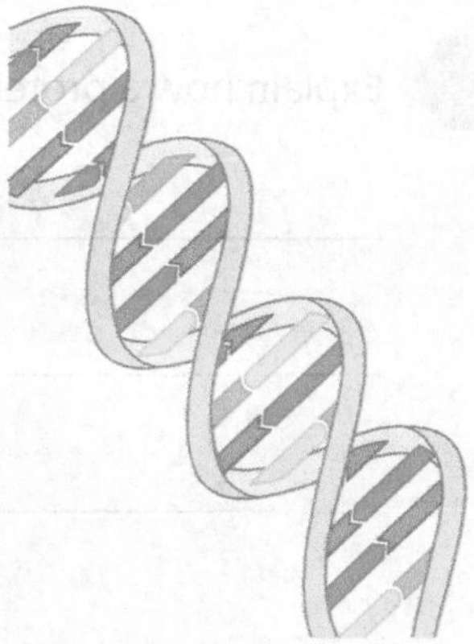

a Name the polymer shown in the diagram.

______________________________ [1 mark]

b Explain why this polymer is essential for life.

______________________________ [1 mark]

4. The diagram shows four nucleotides.

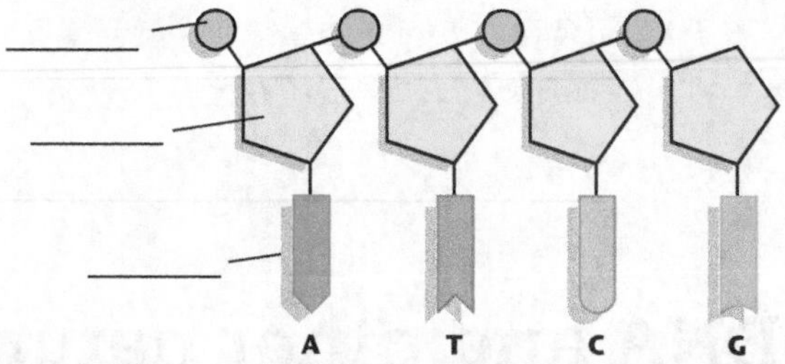

A T C G

a Use words from the box to label the diagram.

acid	base	phosphate	sucrose	sugar	sulfate

[3 marks]

b Circle the nucleotide with the adenine base. [1 mark]

Pure substances, mixtures and formulations

1. Look at the boxes.

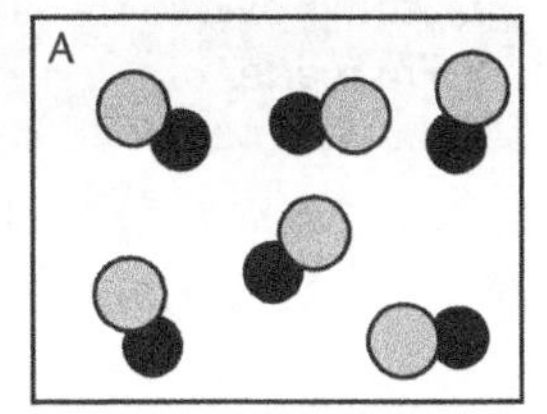

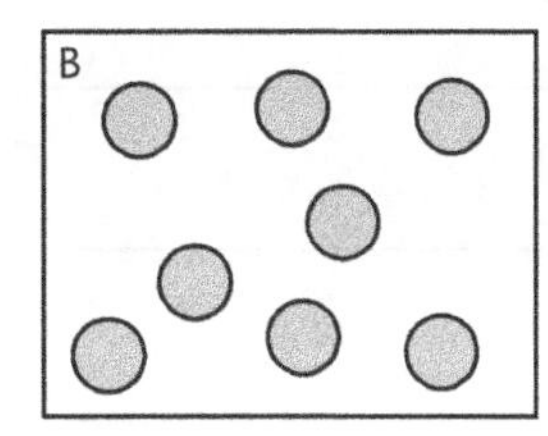

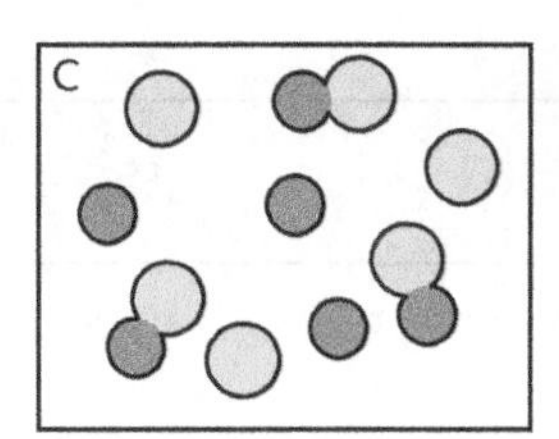

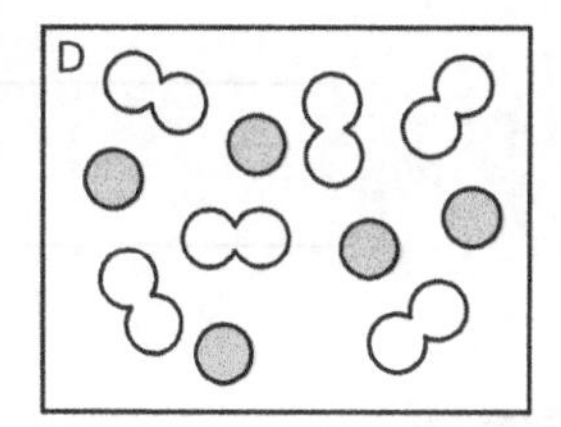

a Which box could be used to represent hydrogen chloride gas? Give a reason for your answer.

Box ____________ Reason: ________________________ [2 marks]

b Which box could be used to represent a mixture of helium gas and hydrogen gas? Give a reason for your answer.

Box ____________ Reason: ________________________

____ [3 marks]

2. Hand cream is a formulation. Explain why substances in hand cream formulations have to be mixed in the correct proportions.

____

____ [2 marks]

3. Look at the food label taken from a carton of orange juice. The manufactures claim that the carton contains pure orange juice. Explain why this claim is incorrect.

____

____

____

____ [3 marks]

PURE ORANGE JUICE

Ingredients: Water, and orange juice concentrates

NUTRITION FACTS Serving Size 8 fl oz (240 ml)	
Amount Per Serving	
Calories 120 Calories from Fat 0	
	% Daily Value*
Total Fat 0g	0%
Sodium 0mg	0%
Potassium 425mg	12%
Total Carbohydrate 29g	10%
Sugar 25g	
Protein 1g	
Vitamin C 110%	• Calcium 2%
Thiamine 15%	• Folate 15%

Here you need to use the word 'pure' in its chemical sense.

4. Explain how you would separate a mixture of sand and salt.

> You should include these key words in your answer: dissolve, filter, evaporate and crystallise. A good answer would contain points made in a logical order, with a reason provided for each point made.

[4 marks]

5. A student made some paracetamol in the laboratory. The melting temperature of the laboratory sample was measured to be 145–155 °C. The melting point of paracetamol in a data book is given as 169 °C. Explain the difference between the two melting points.

[4 marks]

Chromatography and R_f values

1. Chromatography can be used to assess the purity of compounds.

a Look at the chromatogram shown in the diagram above.

Which sample was found to be impure? ____________ [1 mark]

b Describe why the different substances travel difference distances in a chromatogram.

____

____

____ [3 marks]

2. A student used chromatography to investigate the components of food dyes used in sweets. The resulting chromatogram is shown in the diagram.

In this question you will be assessed on using good English, on how you organise your information and on using specialised terms where appropriate.

Literacy **a** Describe what the student needs to do to produce a chromatogram.

____

____

____

____

____

____

____

____

[6 marks]

Use your knowledge of Required practical 6 (Chromatography) to answer this question. Remember to identify the pieces of equipment you use in this experiment.

b Use the chromatogram shown in this question. Describe what the chromatogram tells you about the sample.

____

____

____ [3 marks]

c Describe what the student must do next to identify the green spot.

[1 mark]

d The green spot moved 10.2 cm.

Calculate the R_f value of the green spot, giving your answer to 2 significant figures.

[2 marks]

3. The R_f values for two substances are listed in the table.

Maths

Substance	R_f value	Distance travelled in cm
A	0.86	12.0
B	0.41	

Maths

You will need to recall the equation for R_f.

Look at the diagram to find out the distance that the solvent moved. The question gives you the other piece of data.

a Calculate how far the solvent travelled.

[2 marks]

b Calculate the distance travelled by substance B. Give your answer to 2 significant figures.

[2 marks]

4. A group of students each carried out a chromatography experiment. They recorded their results in the table below.

Experiment	Distance travelled by solvent front (cm)	Distance travelled by blue spot (cm)	Distance travelled by red spot (cm)	Distance travelled by spots in mixture (cm)
1	9.5	4.5	7.1	4.5, 7.2
2	9.6	4.6	7.1	4.5, 7.1
3	9.1	4.1	6.5	4.1, 6.5

a Give the conclusions can you draw from these results.

____

____ [2 marks]

b Why did the students repeat the experiment.

____ [1 mark]

c The measurements for the distance travelled by the spots in experiment 3 are lower than those obtained in experiments 1 and 2. Suggest a reason why.

____ [1 mark]

d When the students were checking their results, one student noticed a dark green spot on the start line where the original mixture had been. Suggest why the spot had not moved during the experiment.

____

____ [2 marks]

e Explain how the student could identify the spot.

____

____

____ [3 marks]

Tests for common gases

1. A student collected three test tubes of gas during an investigation. The student thought that the gas was either carbon dioxide or oxygen. Describe how the student can confirm the identity of the gas.

____

____ [2 marks]

2. When magnesium is mixed with hydrochloric acid, fizzing is observed.

A student wanted to collect a sample of gas so that they could test it. They set up the equipment as shown in the diagram.

Hydrochloric acid

Magnesium

a Identify **two** errors the student made when setting up the equipment.

____

______________________________________ [2 marks]

The student then set up the equipment correctly and collected test tubes of gas. The gas did not bleach damp litmus paper but did make a 'popping' noise when held near a lighted splint.

Worked Example

b Explain what these results show. [4 marks]

The positive test for chlorine gas is that it bleaches damps litmus paper. [1 mark] This gas did not bleach the paper so chlorine gas is not present. [1 mark] The positive test for hydrogen gas is that it burns with a 'pop' when set a light by a lighted splint. [1 mark] The gas does this, so hydrogen is present. [1 mark]

This answer would get full marks because it clearly states what the gases were tested for along with a positive result. The positive results and the actual results are compared and a conclusion is made.

Flame tests

1. Dilute hydrochloric acid was added to a white solid (compound X) and fizzing was observed. When a flame test was carried out, compound X gave a green flame.

You need to learn the flame colours for the metal ions given in the specification.

a What is compound X?

Tick **one** box.

☐ $CuCO_3$ ☐ CuO ☐ K_2CO_3 ☐ Na_2O [1 mark]

b Write a balanced symbol equation for the reaction between compound X and hydrochloric acid. Include state symbols.

______________________________________ [3 marks]

2. Low sodium table salt contains some potassium chloride as well as sodium chloride. Flame tests can be used to identify potassium ions and sodium ions.

This question links to Required practical 7.

a Describe how to carry out a flame test.

____

____ [2 marks]

b Write the colour of the flame you would expect to see if the sample was normal table salt.

____ [1 mark]

c Explain how this result might be the same if the sample tested was a low sodium table salt.

____

____ [2 marks]

Metal hydroxides

1. Some metal ions can be identified by precipitation reactions. Explain what a precipitation reaction is.

____

____ [2 marks]

2. The solution of an ionic compound formed a white precipitate when a small amount of sodium hydroxide was added.

Describe two further tests you could perform to identify the metal ion present. Explain the results you might obtain and give the conclusions you would reach. [4 marks]

Add excess sodium hydroxide to the sample. [1 mark] If the precipitate dissolves, the metal ion is aluminium. If the precipitate does not dissolve the metal ion is calcium or magnesium. [1 mark]

Carry out a flame test. [1 mark] If the flame is orange-red, the metal ion is calcium. [1 mark] If both tests are negative the metal ion is magnesium.

3. **a** Describe what happens when sodium hydroxide solution is added to iron(II) sulfate solution.

[1 mark]

b Write a balanced equation for the reaction between sodium hydroxide solution and iron(II) sulfate solution. Include state symbols.

[3 marks]

4. Aqueous sodium hydroxide was added to a metal chloride solution and a blue precipitate was formed.

a Give the name of the metal ion present. ____________________ [1 mark]

Higher Tier only

b Complete the ionic equation for the reaction.

Cu^{2+} + ____ → $Cu(OH)_2$ [2 marks]

Tests for anions

1. A student carried out some tests on an unknown ionic compound. The results are shown in the table.

Test	Result	Conclusion
Add barium chloride and hydrochloric acid	No change	
Add silver nitrate and dilute nitric acid	Yellow	
Add a little sodium hydroxide solution	White precipitate	
Add excess sodium hydroxide solution	White precipitate	
Flame test	Orange-red flame	

a Write in the table the conclusions you can make for each test. [5 marks]

b Identify the unknown compound. ____________________ [1 mark]

c A student was trying to identify another unknown substance. The student thought it was iron(II) bromide. Describe the tests and results that could confirm this.

____

____

____

____ [4 marks]

Instrumental methods

1. Modern analytical laboratories are equipped with a range of instruments that are used to carry out investigations and identify elements and compounds.

Explain why scientists often choose to use instrumental methods over the more traditional chemical tests.

____

____

____ [3 marks]

2. A mixture of gases was separated using gas–liquid chromatography.

The resulting chromatogram is shown below.

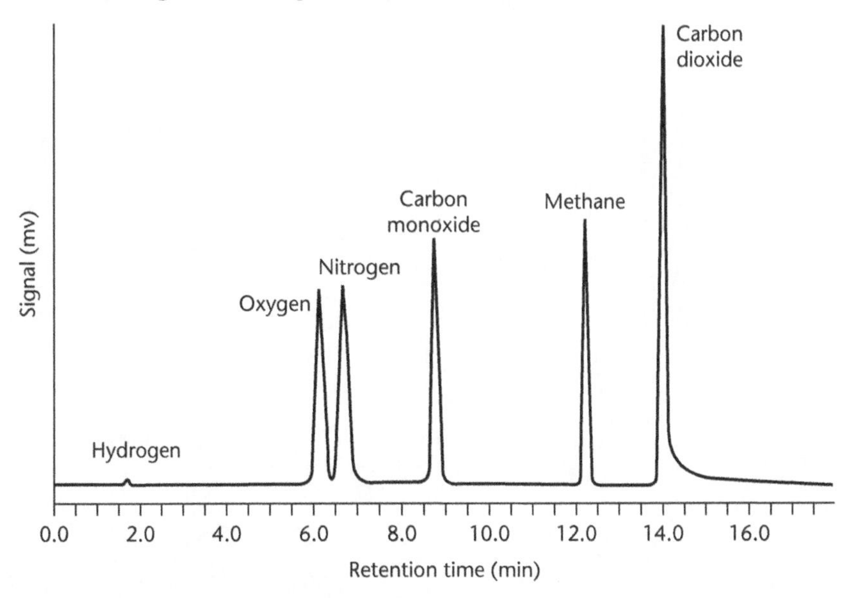

a Give the number of gases present in the mixture. ______ [1 mark]

b Identify the gases that came through between 10 minutes and 16 minutes of the experiment.

______ [1 mark]

c Identify the **least** abundant gas in the mixture. Give a reason for your answer.

______

______ [2 marks]

3. Complete the sentences to describe how you would obtain and use a flame emission spectrum.

The sample is put into a flame and ______

______

The output is a line spectrum that can be analysed to ______

______ [2 marks]

The Earth's atmosphere – now and in the past

1. The pie chart shows the proportion of the gases in the atmosphere today.

C
B
A

a Name gas A in the pie chart. ____________________ [1 mark]

b Segment C in the pie chart represents a mixture of different gases. Name **one** gas in this mixture.

____________________ [1 mark]

2. The proportion of oxygen in the atmosphere can be measured in the laboratory by heating copper metal in a known volume of air.

Dry air
Copper
Heat
Dry air

Re-read the information given at the start of the question.

Volume of dry air in apparatus at the start of the reaction = 60 cm^3

Volume of dry air in the apparatus at the end of the reaction = 48.5 cm^3

a Write a word equation for the chemical reaction that is taking place in the apparatus.

____________________ [1 mark]

Maths **b** Calculate the percentage of oxygen in this sample of air.

Maths
The first step is to work out how much oxygen has been used up in the reaction.

[2 marks]

c The accepted value for the percentage of oxygen in air is approximately 21%. Suggest **two** possible causes for the calculated experimental value being less than 21%.

____

____________________ [2 marks]

3. There are many theories about the early atmosphere.

a Suggest **one** reason why there are different theories about the early atmosphere.

______________________________ [1 mark]

Here is one such theory.

The early atmosphere was made up of gases from volcanoes. These gases included carbon dioxide, nitrogen, ammonia, methane and steam. The gases were released over billions of years. The first oceans formed when the steam condensed as the Earth cooled down. The first oxygen gas was formed by photosynthesis in plants. The amount of carbon dioxide was reduced as it dissolved in the first oceans, forming carbonates. Eventually sedimentary rocks and fossil fuels containing carbon were formed.

b Describe some of the evidence and assumptions scientists based this theory on.

____

____

____

______________________________ [4 marks]

This is a 4-mark question so you should be aiming to include at least two pieces of evidence and two assumptions that scientists would have taken in to account when coming up with this theory. Do not simply just write down pieces of evidence without linking them to an assumption.

Changes in oxygen and carbon dioxide

1. Photosynthesis by algae and plants produced the oxygen that is now in the atmosphere.

a Balance the equation for the reaction.

___ CO_2 + ___ $H_2O \rightarrow C_6H_{12}O_6$ + ___ O_2 [2 marks]

b Explain what happened to the carbon dioxide and oxygen levels in the atmosphere as more plants evolved.

____

____ [2 marks]

c One theory suggests that the oxygen produced by the first algae was removed by the oxidation of iron. The evidence for this is banded iron oxide sediments, which are red, that are nearly always found in older rocks.

If this theory is correct, describe how the levels of oxygen may have changed in the early atmosphere. Suggest a reason for your answer.

____

____

____

____ [4 marks]

2. Explain why the proportions of gases in the atmosphere have been much the same as they are today for the last 200 million years. [3 marks]

Worked Example

During photosynthesis, green plants use carbon dioxide to produce oxygen. [1 mark] During respiration, both plants and animals use oxygen and produce carbon dioxide. [1 mark] These two processes help to keep the proportions of oxygen and carbon dioxide in balance. Nitrogen gas is very unreactive and therefore the amount in the atmosphere has not really changed. [1 mark]

Analysing questions

There are 3 marks available for this question, so you need to talk about the three main gases found in the atmosphere.

Greenhouse gases

1. Greenhouse gases in the atmosphere maintain temperatures on Earth at a level that can support life.

a Which of these gases are greenhouse gases?

Tick **two** boxes.

☐ Carbon dioxide ☐ Methane

☐ Nitrogen ☐ Oxygen [2 marks]

Literacy **b** Describe what is meant by 'the greenhouse effect'.

____

____

____

____ [4 marks]

2. The graph shows the concentration of carbon dioxide in the atmosphere above Hawaii from 2000 to 2015.

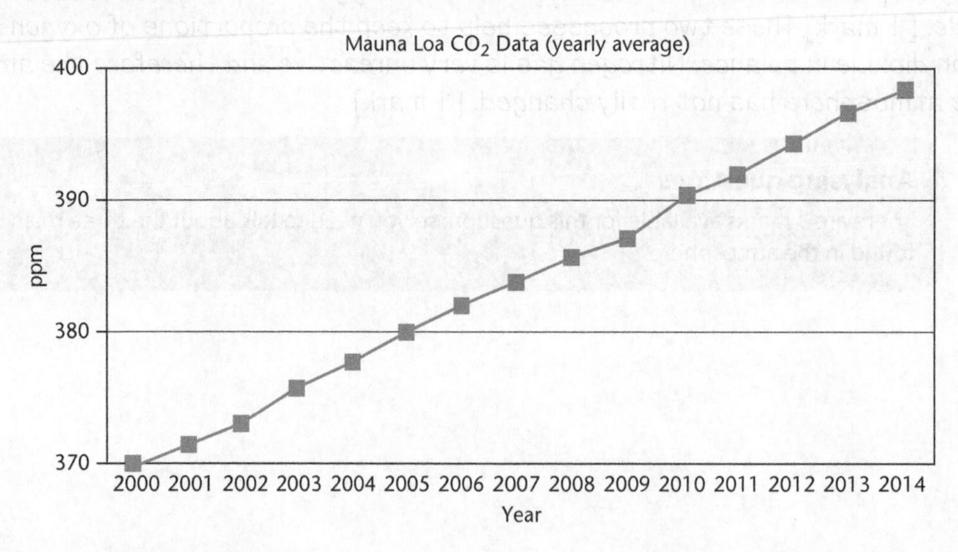

a Write what 'ppm' stands for. ____________________ [1 mark]

Maths **b** Calculate the approximate percentage increase in CO_2 from 2000 to 2014. Give your answer to 3 significant figures.

_______________ [3 marks]

> **Maths**
> You need to use the graph to answer this question.
> 1. Obtain the data from the graph.
> 2. Calculate the increase in ppm with a subtraction sum.
> 3. Now you can calculate the percentage increase.

c Some scientists are really alarmed about the rate at which carbon dioxide has increased during this time but some others are not so worried. Suggest a reason for this difference in opinion.

____

_______________ [2 marks]

d Describe **two** human activities that may have caused the increase in CO_2 levels.

____

_______________ [2 marks]

Global climate change

1. Scientist use a number of different indicators to demonstrate how the climate is changing. Which of the following is **not** an indicator of climate change? Tick **one** box.

- ☐ Changes in weather systems
- ☐ Increase in agriculture
- ☐ Changes in migration of birds
- ☐ Retreat of glaciers

[1 mark]

2. Explain how scientists can use glaciers to demonstrate how the climate is changing.

____

____

_______________ [3 marks]

3. If global warming continues there will be big changes in the environment.

Literacy Describe environmental changes that could take place. Evaluate the possible consequences and impacts.

____

____

____

____

____

______________________________ [6 marks]

Here you need to state how the climate could change and discuss how those changes could affect both humans and other living species.

Carbon footprint and its reduction

1. State the meaning of the phrase 'carbon footprint'.

____

______________________________ [1 mark]

2. **a** It is decided to fit solar panels to all the new houses in a development. Explain how this will help reduce CO_2 emissions.

____

______________________________ [2 marks]

b Explain the disadvantages of using solar panels as an alternative energy source.

> You need to consider how much energy is required during manufacture and where that energy comes from.

____

____

____ [3 marks]

3. The graph shows the greenhouse gas (GHG) emissions associated with UK consumption for the years 1997 to 2013.

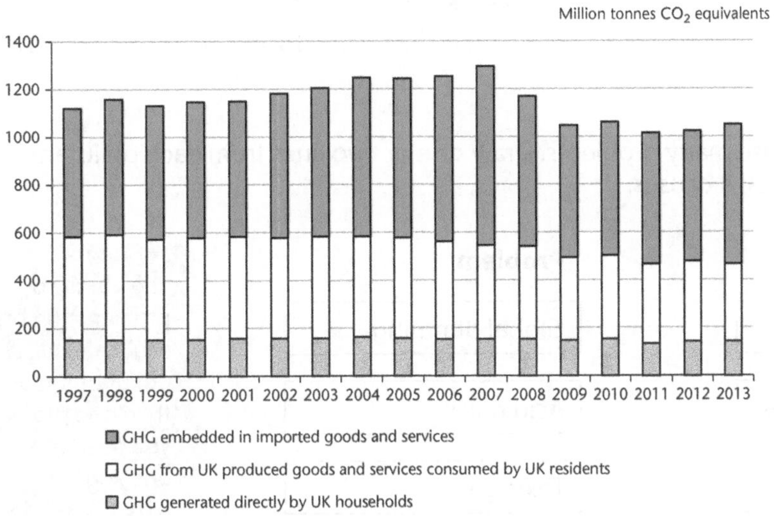

Million tonnes CO_2 equivalents

1400
1200
1000
800
600
400
200
0

1997 1998 1999 2000 2001 2002 2003 2004 2005 2006 2007 2008 2009 2010 2011 2012 2013

- GHG embedded in imported goods and services
- GHG from UK produced goods and services consumed by UK residents
- GHG generated directly by UK households

a Give the year when the carbon footprint reached a peak. Suggest a reason for your answer.

____

____ [2 marks]

b Name the sector that has the highest carbon footprint. Suggest a reason for your answer.

____

____ [2 marks]

c Describe the trend seen in the carbon footprint generated directly by UK households.

____ [1 mark]

Literacy **d** Explain how individual households can reduce their carbon footprint.

> Your answer needs to give a reason how/why each method you have chosen will help to reduce the overall carbon footprint.

[4 marks]

Air pollution from burning fuels

1. Air pollution can cause many problems. Draw **one** or **two** lines from each pollutant to the problem (or problems) it causes.

Pollutant	Problem
Carbon monoxide	Global dimming
Carbon particulates	Acid rain
Sulfur dioxide	Toxic gas
Nitrogen dioxide	Respiratory problems

> The phrase 'or problems' (in brackets) suggests that some pollutants cause more than one problem and you need more than one line.

[4 marks]

2. The quality of our air can be improved by reducing amount of pollutants we emit. Draw lines from each pollutant to a method of removal.

Pollutant	Removed by
Carbon monoxide	Catalytic converter
Sulfur dioxide	Capturing with limestone
Nitrogen dioxide	

[3 marks]

3. Nitrogen dioxide (NO_2) and carbon dioxide are pollutants often formed by car engines.

a Describe the conditions required for nitrogen dioxide to form.

________________________________ [1 mark]

b Write a balanced equation, including state symbols, for the formation of nitrogen dioxide.

________________________________ [3 marks]

Count the atoms on both sides of the equation to balance it. You can only change the number of molecules not the number of individual atoms in a molecule.

c Heptane is used as the fuel in some internal combustion engines. Heptane burns in different ways, which can be illustrated by these two equations.

Reaction A: $C_7H_{16} + 8O_2 \rightarrow CO_2 + 6CO + 8H_2O$

Reaction B: $C_7H_{16} + 11O_2 \rightarrow 7CO_2 + 8H_2O$

Compare reaction A with reaction B and use them to explain why carbon monoxide is sometimes formed.

Command words

'Compare' means to look for similarities *and* differences.

____

____

____

________________________________ [4 marks]

d Explain why most cars are fitted with catalytic converters.

________________________________ [1 mark]

What does the Earth provide for us?

1. Sustainable development is about meeting people's present needs without spoiling the environment for the future.

Give an example of sustainable development and explain your answer.

[2 marks]

2. Crude oil is an example of a finite resource. Wood is an example of a renewable resource. Compare the two types of resource.

[3 marks]

3. Graph A shows the water footprints of different foodstuffs and graph B shows the change in meat consumption per person in two different countries.

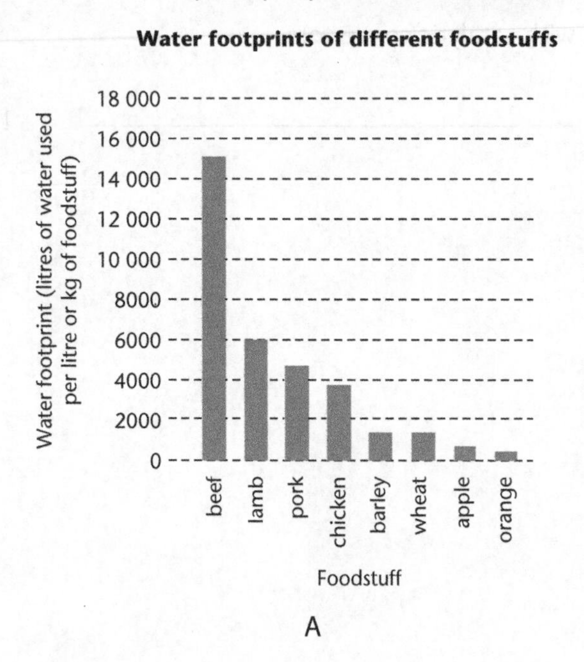

A

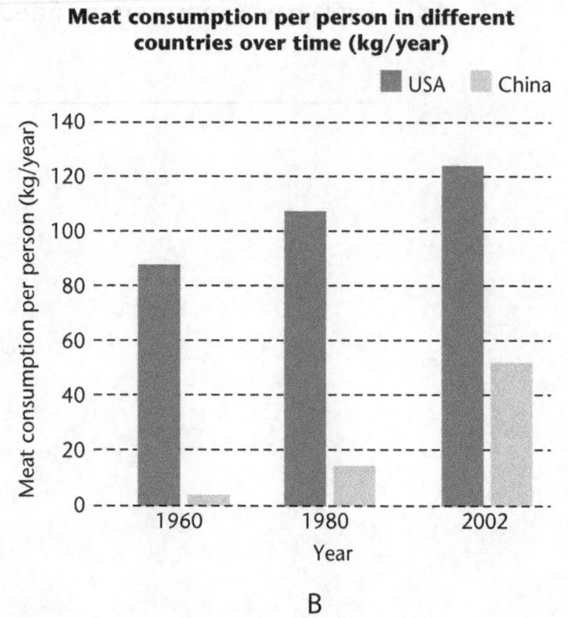

B

Use the data provided to show how changing diets around the world may be linked to sustainable development.

____

____

____

____

____

______________________________ [6 marks]

Safe drinking water

1. In many countries, such as the United Kingdom, fresh water is supplied from reservoirs.

a Describe where the fresh water in the reservoirs comes from.

______________________________ [1 mark]

b The three main stages in water treatment are sedimentation, filtration and sterilisation. Give a reason for each stage.

____

____

______________________________ [3 marks]

2. Some countries in the Middle East obtain their drinking water from seawater.

Required practical

a Describe how distillation can be used to obtain drinking water from seawater.

____

____

______________________________ [3 marks]

b Explain why distillation is not usually used to purify drinking water.

[1 mark]

c Explain the difference between pure water and potable water.

[3 marks]

3. Many people filter their tap water because it removes impurities and improves the taste.

Required practical

Describe a test that could be carried out to determine the purity of the filtered water.

[2 marks]

Waste water treatment

1. People living in cities produce large amounts of waste water. Suggest why the waste water needs treatment before being released into the environment.

As this question has 2 marks, you should try to write down two suggestions.

[2 marks]

2. Sewage treatment includes both the **anaerobic** digestion of sewage sludge and the **aerobic** biological treatment of effluent. Explain the difference in meaning between 'anaerobic' and 'aerobic'.

[1 mark]

3. Modern lifestyles require organised waste water management systems.

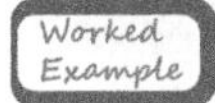

Compare the processing of sewage, agricultural waste water and industrial waste water. [3 marks]

Sewage and agricultural waste water always require the removal of organic waste whereas industrial waste water may not. [1 mark] Sewage and agricultural waste water contain harmful microbes which need to be removed but industrial waste water does not. [1 mark] Instead it contains harmful chemicals that need to be removed. [1 mark]

> Suitable connectives for a 'compare' question include 'whereas', 'but', 'instead of' and 'otherwise'.

Alternative methods of extracting metals

1. Useful metals, such as copper, have been mined from ancient times but now alternative methods of extraction are being developed.

Higher Tier only

a Give two reasons why new methods of extraction are being developed.

____

____ [2 marks]

b Use words from the box to complete the sentences.

bioleaching	burned	electrolysed	harvested
metal	phytomining	processed	

____________ uses plants to absorb ____________ compounds.

The plants are grown, ____________ and then ____________

to produce ash that contains metal compounds. [4 marks]

2. Copper can be extracted by an alternative method that uses plants.

a Name the copper compound found in the ash after the plant has been burnt. ____________ [1 mark]

Worked Example

b Describe how copper metal can be obtained from copper oxide using scrap iron. [3 marks]

Step 1: React copper oxide with dilute sulfuric acid to make copper sulfate solution. [1 mark]

Step 2: Add the iron to the copper sulfate solution. A displacement reaction will take place leaving the copper. [1 mark]

> This answer is correct but in step 2 you could use obtain the copper by electrolysis instead of using a displacement reaction.

Step 3: Filter the solution. The copper metal will be the residue left in the filter, which can be dried and removed. [1 mark]

3. Suggest why it still may be more economically viable to mine ores it the traditional way than to use alternative biological methods.

This question is just about economic arguments but you also could be asked about environmental impacts.

____

____

______________________________ [3 marks]

4. Explain why and how alternative methods of extracting metals such as zinc, cobalt or nickel are now being used instead of more traditional methods.

Literacy

____

____

____

____

____

____

____

______________________________ [6 marks]

Life cycle assessment (LCA)

1. Which of these gives the four stages of life cycle assessment in the correct order? Tick **one** box.

- ☐ Development, extracting and processing raw materials, use, disposal
- ☐ Extracting and processing raw materials, manufacture, use, disposal
- ☐ Manufacture, desirability, use, disposal
- ☐ Manufacture, use, cost, disposal

[1 mark]

2. The table contains some life cycle assessment data for different types of bags. The data can be used to compare the sustainability of different products.

	Paper bag	Plastic bag
Raw materials	Wood	Crude oil
Energy used during production of bag	1.8 MJ	1.5 MJ
Solid waste	52 g	16 g
Total emissions to air	2.8 kg	1.3 kg
Greenhouse gases (CO_2 equivalents) produced	0.23 kg	0.53 kg

a Look at the information in the table. Determine the type of bag that is better for the environment. You must give reasons for your answer.

____

____

____

______________________________ [4 marks]

b Suggest why it is important to identify who has made a life cycle assessment.

______________________________ [1 mark]

3. *Carriage Coffee Express* is about to open a new café at a busy railway station. One of the final jobs for the new manager is to decide what type of coffee cup to use. The manager consults some data that summarises the environmental impact of each type of coffee cup.

	Human health	Quality of ecosystems	Climate change	Resource depletion	Water consumption
Ceramic mug	Good	Good	Good	Good	High
Travel mug	Average	Poor	Good	Good	Low
Paper cup (PE)	Poor	Average	Poor	Poor	Low

Note that the data for both ceramic mugs and travel mugs assumes 500 uses.
Use the data in the table to help the manager to decide what type of cup to use.

____

____

____

____ [4 marks]

Ways of reducing the use of resources

1. A large percentage of UK households still do not recycle enough of their waste. As a result, we throw lots of things away. Use the information in the diagram to answer these questions.

a Give the percentage of the waste that can be composted.

____________________ [1 mark]

b List two materials that could be recycled.

____

____________________ [2 marks]

c Explain what the owner of this waste bin could do to reduce household waste.

____

____ [2 marks]

2. The phrase 'reduce, reuse, recycle' is often used when thinking about the environment. Explain what the phrase means, using examples.

The final two words 'using examples' are important in this question.

____

____

____ [3 marks]

3. The graph shows the household waste recycling rate in England from 2000 to 2017.

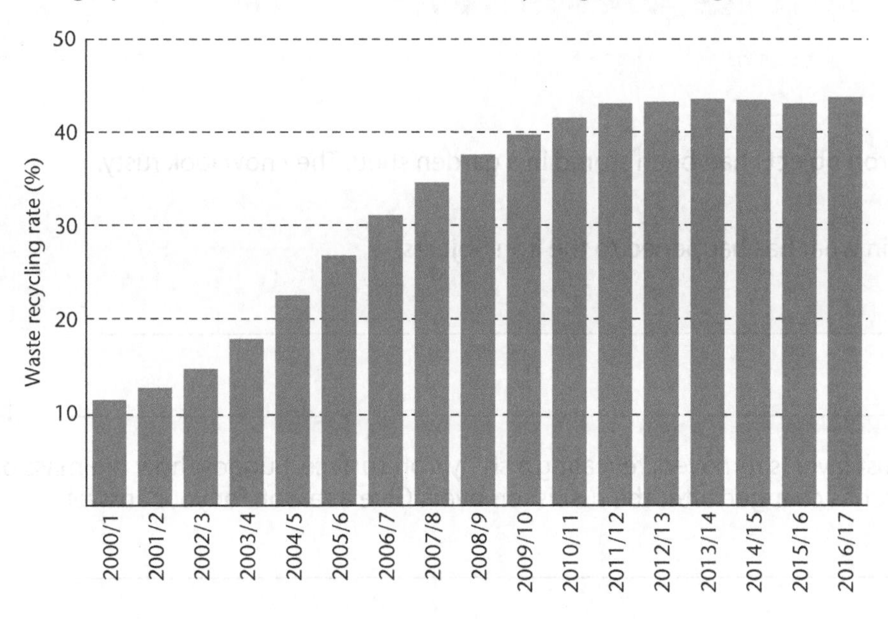

a Describe how the household recycling rate has changed between 2000 and 2017.

____

____

____

____ [3 marks]

b Suggest **two** reasons why the government wants to encourage household recycling.

____

____ [2 marks]

Literacy **c** Explain how pollution problems can be reduced by recycling materials.

____

____

____

____ [4 marks]

Corrosion and its prevention

1. Some cast iron objects had been stored in a garden shed. They now look rusty.

a Explain what has happened to the iron objects.

____

____ [2 marks]

b The rust layer is removed, revealing a shiny iron surface. Suggest how the mass of the object has changed after the rust is removed. Give a reason for your answer.

____

____ [2 marks]

c Describe how the objects could be protected from further rusting.

____ [1 mark]

2. Explain why drinks cans are often made from aluminium, not iron.

____

____

____

____ [4 marks]

3. A block of magnesium metal or zinc metal is often attached to hull of a ship to prevent the ship's hull from rusting. The process is called sacrificial protection.

During the initial stages of rusting the iron is oxidised and the oxygen is reduced.

a Complete the ionic equation that shows the oxidation of iron.

$Fe \rightarrow Fe^{2+} +$ ____________________ [1 mark]

b Use the ionic equation below to show that the oxygen is reduced.

$O_2 + 4e^- + 2H_2O \rightarrow 4OH^-$

____ [1 mark]

c Explain why you might choose to use magnesium rather than zinc to prevent the hull of a ship from rusting.

[1 mark]

d Explain how sacrificial protection works. Include equations in your answer.

[5 marks]

Alloys as useful materials

1. Brass is an alloy.

a Describe what an alloy is.

[1 mark]

b The diagram shows a particle model of brass.

Zinc

Copper

Calculate the percentage of copper atoms present in this sample of brass.

[2 marks]

2. The table gives some useful information about some pure metals and some alloys.

Metal or alloy	Density (g/cm³)	Melting point (°C)	Electrical resistance (10^{-8} Ω m)	Tensile strength (MPa)
Aluminium	2.7	934	2.6	80
Aluminium alloy	2.8	800	5	600
Iron	7.8	1810	10	300
Steel	7.8	1700	15	460

a Use data from the table to explain why an engineer would choose to use aluminium alloy when designing a helicopter.

Your answer must be based on information from the table in the question; otherwise no marks can be awarded. For each point you make you must give a reason why the engineer would make this choice.

____

____

____

______________________________ [4 marks]

b An engineer needs some metal rods to reinforce a concrete pillar. Explain why steel is a better choice for this than iron.

____

______________________________ [2 marks]

3. Gold used for jewellery is often alloyed with other metals such as copper, silver or zinc. The proportion of gold in each alloy is measured in carats. The term '24 carat gold' means the sample is 100% gold.

a Explain why 18 carat earrings will be more expensive than 6 carat earrings of the same size and design.

____

______________________________ [2 marks]

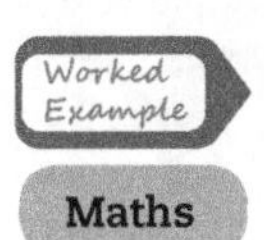

Maths

b Calculate the mass of gold used to make an 18 carat earring that has a mass of 3 g. [2 marks]

fraction of gold in each earring = $\frac{18}{24}$ = 0.75
[1 mark]

mass of gold in an earring = 0.75 × 3 = 2.25 g
[1 mark]

This answer is correct and clearly shows all the working and uses the correct units.

1 mark is awarded for working out the fraction of gold, using the definition of carat in the question opener, and the other mark is awarded for working out the actual mass of gold.

Ceramics, polymers and composites

1. The table shows some properties of some materials.

Material	Density (g/cm³)	Melting point (°C)
Low density poly(ethene) (LDPE)	0.90–0.93	105–115
High density poly(ethene) (HDPE)	0.93–0.98	120–180
Wood	0.6–0.9	Burns when gets hot
Concrete	2.4	Decomposes between 500 and 950 °C
Ceramic	1.7–2.0	>1800

a Identify a composite material from the table. Give a reason for your answer.

____ [2 marks]

b Bricks are ceramic materials made by baking clay. Explain why kilns are often lined with bricks.

____ [1 mark]

c Explain how and why the properties of LDPE and HDPE differ.

[4 marks]

2. Literacy Compare the structure and bonding in a thermosoftening polymer, such as poly(ethene), with the structure and bonding in a thermosetting polymer, such as epoxy resin.

[6 marks]

The Haber process

1. Ammonia is made in the Haber process when nitrogen and hydrogen are passed over an iron catalyst.

a Balance the equation for the Haber process reaction. Add state symbols.

____ N_2 ____ + ____ H_2 ____ $\rightleftharpoons$ ____ NH_3 ____ [3 marks]

b State the meaning of the symbol $\rightleftharpoons$.

[1 mark]

c Describe what will happen if the pressure in the Haber process is increased.

____

____ [2 marks]

Higher Tier only

d Calculate the mass, in tonnes, of ammonia which could be produced from 375 tonnes of nitrogen.

Relative atomic masses, A_r: H = 1; N = 14

Give your answer to 3 significant figures. [3 marks]

You will need to recall and then use the equation:

$$\text{number of moles} = \frac{\text{mass of sample}}{\text{molecular mass}}$$

and remember that the molecular mass must be worked out in grams.

1 tonne = 1 000 000 g or 1×10^6 g. When using large numbers, it is often easier to work in standard form.

2. The graph shows how the percentage of ammonia produced in the Haber process changes with temperature and pressure.

$N_2(g) + 3H_2(g) \rightleftharpoons 2NH_3(g)$

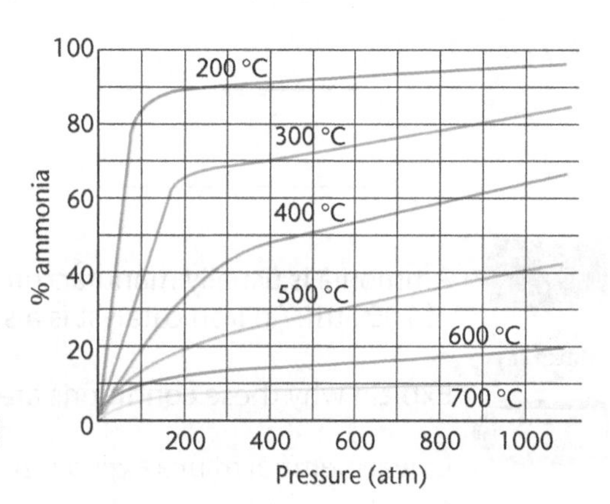

a Estimate the percentage of ammonia formed at 300 °C and 400 atm pressure.

______________________________ [1 mark]

To find the value you must draw a line up from 400 atm to the blue line labelled 300 °C and then draw a second line to the *y*-axis.

b Describe how the percentage of ammonia formed changes at 400 atm as the temperature increases. Suggest a reason for your answer.

> Use the graph to answer the first part of the question. Then apply your knowledge and understanding of equilibria (see Section 5) to suggest a reason for the observation.

[3 marks]

c Suggest reasons why the Haber process reaction is not carried out a room temperature and pressure.

[2 marks]

d Explain why the nitrogen and hydrogen are passed over iron in the industrial manufacture of ammonia.

[2 marks]

3. Literacy

Ammonia is usually manufactured using a temperature of 400–450 °C and a pressure of 200 atm. An iron catalyst is also used.

Explain why these conditions are chosen. [4 marks]

Worked Example

Lower temperatures give higher yields but the reaction rate is very slow, so higher temperatures are used. An iron catalyst is used to increase in the rate of reaction. [1 mark] Working at very high pressures gives a high yield but the equipment is very expensive. [1 mark] The chosen conditions give a yield of about 15% ammonia but the reactants are recycled through a continuous flow system. [1 mark] To make the process viable we need a compromise between position of equilibrium, rate of reaction and cost. [1 mark]

> This answer would gain full marks.
>
> You should look back at the section on equilibrium to remind yourself of Le Chatelier's Principle, which is applied here.

Production and uses of NPK fertilisers

1. A fertiliser has an N:P:K ratio of 24:6:6. Explain what these numbers mean.

[3 marks]

2. Farmers use a range of different fertilisers on their farms, including ammonium nitrate (NH_4NO_3), potassium sulfate (K_2SO_4) and ammonium phosphate $(NH_4)_3PO_4$.

Explain why a range of fertilisers are used.

[4 marks]

3. Describe how potassium chloride is made in the laboratory. Write an equation for the reaction.

Make sure you write each step in a logical order so that someone can follow your method.

[4 marks]

4. The diagram shows stages in the industrial manufacture of the fertiliser ammonium nitrate, NH_4NO_3.

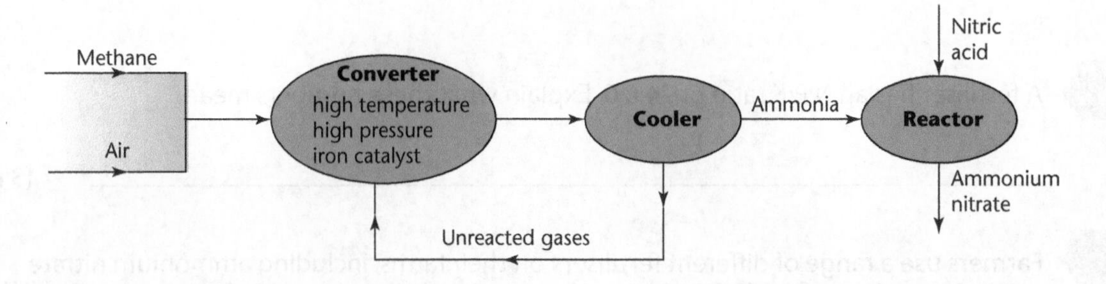

a Write down the word equation for the reaction taking place in the reactor during the production of ammonium nitrate.

____ [2 marks]

b Describe what is happening in the converter.

____

____ [2 marks]

c Ammonium nitrate is produced on an industrial scale by a continuous flow method. In the laboratory it is produced in batches.

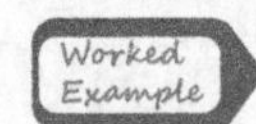

Compare the two methods. [4 marks]

> To get full marks in this type of question, the answer must describe the similarities and differences of both processes.

Continuous flow means that ammonium nitrate is being produced all the time until the production plant is shut down. [1 mark] In a batch process, a known amount of ammonium nitrate is made at a time in one reaction. [1 mark] In continuous flow, as soon as the product is made it is piped away so that more reactants can continue to react. [1 mark] In a batch process, known quantities of reactants are measured out and mixed together. Once the reaction is complete, the product is removed and the process is repeated. [1 mark]

Notes

Notes

Section 1: Atomic structure and the periodic table

Atoms, elements and compounds

1. Carbon dioxide [1 mark]

   Water [1 mark]

2. An element is made up of only one type of atom. [1 mark]

3. **a** Na; [1 mark] Cl_2 [1 mark]

   b Sodium chloride [1 mark]

Mixtures

1. 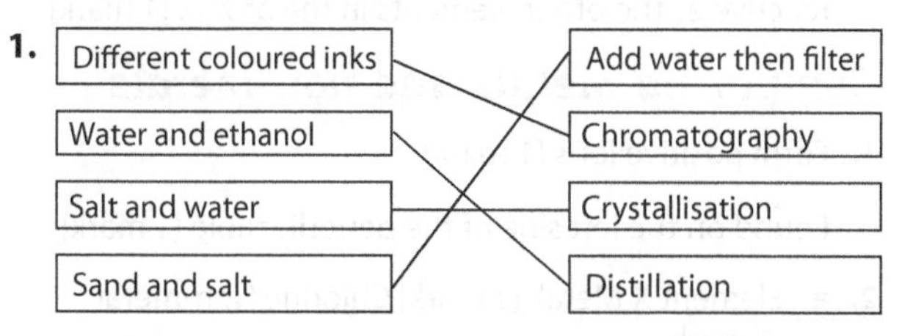

   3 marks for all correct, 2 marks for 2 correct, 1 mark for 1 correct

   Award no mark if more than one line is drawn from/to a box

2. **a** To stop the mixture coming out of the test tube and burning anyone. [1 mark]

   b A [1 mark]

   c Use a magnet. [1 mark]

   d The iron powder will stick to the magnet because it is magnetic. [1 mark] The sulfur powder will be left behind because it is non-magnetic. [1 mark]

3. **a** Distillation [1 mark]

   b Heat the mixture to a temperature of 65 °C. [1 mark] Methanol will boil and leave the mixture as a gas. [1 mark] Condense the gas to form liquid methanol. [1 mark] The ethanol will be left behind. [1 mark]

Compounds, formulae and equations

1. $C_6H_{12}O_6$ 1 mark for correct numbers, 1 mark for numbers as subscripts

2. **a** Lithium chloride [1 mark]

   b Sodium carbonate [1 mark]

   c Barium sulfate [1 mark]

   d Calcium hydroxide [1 mark]

3. magnesium + oxygen → magnesium oxide [1 mark]

   $2Mg(s) + O_2(g) \rightarrow 2MgO(s)$ [1 mark for chemical symbols, 1 mark for balancing, 1 mark for state symbols]

Scientific models of the atom

1. X Neutron; [1 mark] Y Proton; [1 mark] Z Electron [1 mark]

2.

Marks awarded for this type of answer are determined by the Quality of Written Communication (QWC) as well as the standard of your scientific response.
0 marks: No relevant content.
Level 1 (1–2 marks): One or more key points have been given.
Level 2 (3–4 marks): Student can describe the main points of each model.
Level 3 (5–6 marks): Student describes the models using correct terminology. They compare several structures within the models.
Indicative content • In the plum pudding model the atom is a ball of positive charge. • In the nuclear model the positive charge is only in the centre/nucleus. • In the nuclear model positive charge consists of particles called protons. • In the plum pudding model negative charge is spread over the whole atom. • In the nuclear model negative charge/electrons are confined to energy levels/shells. • The plum pudding model does not contain neutrons but the nuclear model does.

3. New evidence was collected [1 mark] by scientists carrying out further experiments. [1 mark] The results showed the (plum pudding) model was not correct. [1 mark] The model had to change to explain the results. [1 mark] The new experiment was the alpha particle scattering experiment/Rutherford experiment/gold leaf experiment. [1 mark]

Sizes of atoms and molecules

1. All atoms have a nucleus. [1 mark]

   Most of the mass of an atom is in the nucleus. [1 mark]

2. **a** 3.5×1000; [1 mark] 3500 mm [1 mark]

   Award 2 marks for correct answer with no working.

   b 1 m = 1000 mm; 1 mm = 1000 μm

   $2 \times 1000 \times 1000$; [1 mark] $= 2\,000\,000 = 2 \times 10^6$ μm [1 mark]

   Award 2 marks for correct answer with no working.

3. **a** 1×10^{-10} m [2 marks]

   If answer is correct but not in standard form, award 1 mark.

   b $\frac{0.1 \text{ nm}}{10\,000} = 0.000\,01$ nm; [1 mark] $= 1 \times 10^{-5}$ nm [1 mark]

Relative masses and charges of subatomic particles

1. Electron: very small; [1 mark] –1 [1 mark]

   Neutron: 0 [1 mark]

   Proton: 1 [1 mark]

2. a An atom contains the same number of protons and electrons; [1 mark] so the negative and positive charges cancel each other out. [1 mark]

   b 7 [1 mark]

   c 3 [1 mark]

   d Lithium [1 mark]

3. The two atoms both contain the same number of protons/have the same atomic number. [1 mark] But they have different numbers of neutrons/different mass numbers. [1 mark]

Relative atomic mass

1. a 35 [1 mark]

   b $^{79}_{35}Br$ 44; [1 mark] $^{81}_{35}Br$ 46 [1 mark]

   c There is (roughly) 50% of each isotope. [1 mark] The relative atomic mass is the mean/average of the isotopes. [1 mark]

2. $92 \times 28 = 2576$; $5 \times 29 = 145$; $3 \times 30 = 90$ [1 mark]

   $\frac{2576 + 145 + 90}{100}$ [1 mark] $= 28.1$ [1 mark]

Electronic structure

1. a 2 electrons on innermost energy level; [1 mark] 8 electrons on second energy level; [1 mark] 1 electron on third energy level [1 mark]

   b Charge of (+)1. [1 mark] A sodium ion has one more proton than it has electrons/it has 11 protons and 10 electrons. [1 mark] Protons have a positive charge and electrons have a negative charge. [1 mark]

2. a 2,8 [1 mark]

   b –2 [1 mark]

Electronic structure and the periodic table

1. atomic; [1 mark] group [1 mark]

2. a 6 [1 mark]

   b Y [1 mark]

   c Element Z has a full outer shell of electrons. [1 mark] Element Z does not need to react to achieve a full outer shell. [1 mark]

   d Elements W and Y are in the same group; [1 mark] so carry out similar chemical reactions. [1 mark]

Development of the periodic table

1. The atomic number of the elements was not known; [1 mark] because protons/electrons had not been discovered at the time he was working. [1 mark]

2. a He left gaps where the next known element did not follow the pattern but would follow the pattern if it was moved one space. [1 mark] He left spaces for elements that had not yet been discovered. [1 mark]

   b So iodine could be placed in Group 7, where it shares properties with other elements (F, Cl and Br). [1 mark]

3. Most elements in the group have similar properties but hydrogen is a gas, not a metal. [1 mark] Copper and silver are placed in Group 1 but they are not as reactive as the other elements in the group. [1 mark]

Comparing metals and non-metals

1. Form positive ions [1 mark]

   Found on the left side of the periodic table [1 mark]

2. a Element X Metal; [1 mark] Chlorine Non-metal [1 mark]

   b Group 3; [1 mark] It forms 3+ ions [1 mark]

3. Potassium floats on water so must have a low density [1 mark]. Most metals have a high density. [1 mark]

   Graphite conducts electricity [1 mark]. Most non-metals do not conduct electricity. [1 mark]

Elements in Group 0

1. They exist as single atoms. [1 mark]

2. a Argon [1 mark]

   b It has a full outer shell of electrons; [1 mark] so does not have to lose or gain any electrons to gain a full outer shell; [1 mark] so does not react/form molecules with other elements. [1 mark]

3. a As the atomic number increases the boiling point increases. [1 mark]

   b Cool the mixture until it reaches –246.1 °C. [1 mark] Neon will turn into a liquid and can be removed. [1 mark] Helium will stay as a gas. [1 mark]

Elements in Group 1

1. a $2Na + 2H_2O \rightarrow 2NaOH + H_2$

   Na; [1 mark] H_2; [1 mark]; correct balancing [1 mark]

   b Any one from: The sodium will float. The sodium will melt/form a ball. The sodium will move around the surface of the water. Fizzing/effervescence bubbles. [1 mark]

2. a All points plotted correctly. [2 marks] Award only 1 mark if one point plotted incorrectly. Line of best fit drawn. [1 mark]

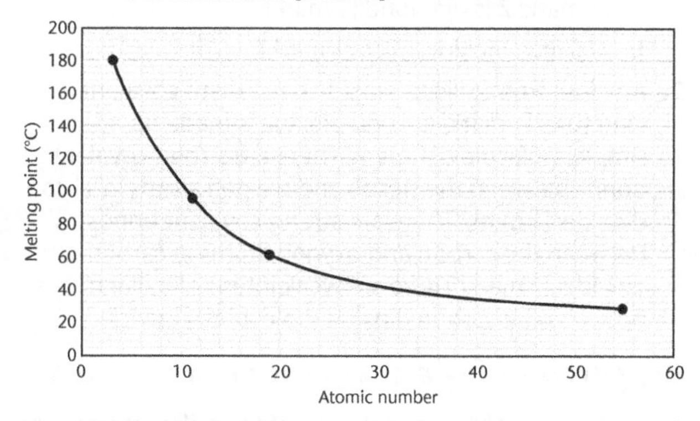

b Answer should be taken from the line of best fit (approximately 38 °C). [1 mark]

Elements in Group 7

1. Chlorine molecules are Cl_2. 35.5 × 2; [1 mark] = 71 [1 mark]
2. a They all have 7 electrons on their outer shell. [1 mark] They will react to gain or share one electron to fill their outer shell which confers stability. [1 mark]
   b $2K(s) + Br_2(l)$ [1 mark] $\rightarrow 2KBr(s)$ [1 mark]
   c The Group 7 elements get less reactive as you go down the group. [1 mark]
3. a $2Br^-(aq) + Cl_2(g) \rightarrow 2Cl^-(aq) + Br_2(aq)$ [2 marks]
   b The clear colourless solution [1 mark] would turn red/brown. [1 mark] More reactive chlorine, which is colourless, has displaced the less reactive bromine from the compound. The bromine is red/brown. [1 mark]

Properties of the transition metals

1. a Nickel [1 mark]
   b +2 [1 mark] The sulfate ion is 2–, the compound is neutral and the ions are in a 1:1 ratio. [1 mark]
   c Any one from: catalyst/dye/pigment/electroplating [1 mark]
2. a Any two from: High melting point/malleable/good electrical conductivity/coloured/high density [2 marks]
   b Group 1 metals are more reactive than transition metals. [1 mark] Accept an example, e.g. Group 1 metals react quickly with water/oxygen/halogens but transition metals react slowly (e.g. zinc/iron) or not at all (e.g. gold) with oxygen/water.
3. a It is strong/hard/does not corrode/is biocompatible. [1 mark]
   b Lithium is reactive; [1 mark] will react with water inside the body and wear away/corrode. [1 mark] Lithium is a soft metal/not very hard/strong; [1 mark] so will break under force. [1 mark]

Section 2: Bonding, structure and the properties of matter

The three states of matter

1. a X [1 mark]
   b Solid [1 mark]
   c 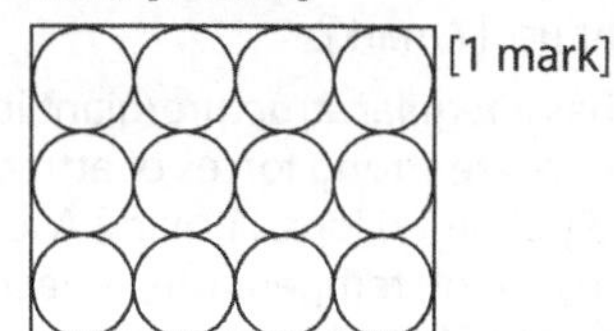 [1 mark]
   d

Marks awarded for this type of answer are determined by the Quality of Written Communication (QWC) as well as the standard of your scientific response.
0 marks: No relevant content.
Level 1 (1–2 marks): Student mentions one argument for or against.
Level 2 (3–4 marks): Student mentions at least one argument for and against.
Level 3 (5–6 marks): Student mentions more than one argument for and against the model.
Indicative content Arguments for: • It shows how the particles are arranged. • It shows that the particles are identical. • It shows that the particles have a regular arrangement. Arguments against: • There are no forces shown between the particles. • It does not show the movement/vibration of the particles. • All particles are represented as spheres. • The model is just two-dimensional.

Ionic bonding and ionic compounds

1. $Ca(NO_3)_2$ [1 mark]
2. b A
   c In A there are 8 of the large spheres and 14 of the smaller ones. [1 mark] This is approximately a ratio of 1:2, so it must contain 1 calcium ion to every 2 chloride ions. [1 mark] In B there are approximately equal numbers of the two spheres, so the ratio is 1:1.

Dot and cross diagrams for ionic compounds

1. b +1 (or 1+) [1 mark]
   c It has one more proton than electrons; protons have a +1 charge. [1 mark]
2. Outer shell of electrons on chlorine atom drawn correctly, with 7 electrons. [1 mark] One electron transferred from sodium atom to chlorine atom. [1 mark] Sodium ion has +1 charge, chloride ion has –1 charge. [1 mark] Name of compound formed is sodium chloride. [1 mark] Formula is NaCl. [1 mark]

Properties of ionic compounds

1. **a** $PbBr_2$ [1 mark]
   b When the lead bromide is solid the ions cannot move and conduct electricity and so the circuit is not complete and the bulb does not light up. [1 mark] After a while the lead bromide melts. [1 mark] The ions are free to move in the liquid lead bromide, the circuit is electrically complete and the bulb does light up. [1 mark]
   c Lead(II) bromide has a regular structure/giant ionic lattice. [1 mark] There are strong forces of attraction between oppositely charged ions. [1 mark] A lot of energy, provided by a high temperature, is needed to overcome the forces. [1 mark]

Covalent bonding in small molecules

1. **a** CO_2 [1 mark] **b** $CaCl_2$ [1 mark]
   c Cl_2 [1 mark] **d** CO_2 [1 mark] Cl_2 [1 mark]
2. 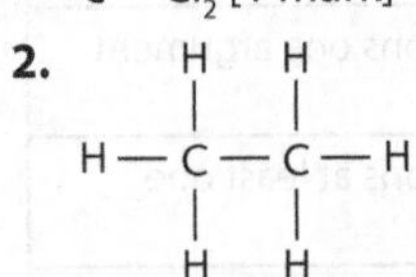

   Line between the two carbons; [1 mark] line from each carbon atom to a hydrogen atom. [1 mark]

Dot and cross diagrams for covalent compounds

1. 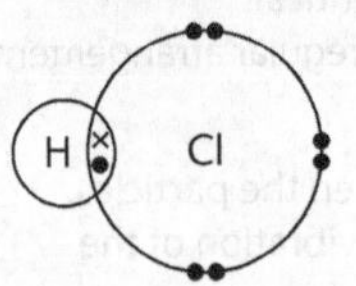

   One atom of each element; [1 mark] outer shell of atoms are overlapped; [1 mark] one electron from each element in the overlapped space. [1 mark]
2. Six electrons on outer shell of oxygen atoms; [1 mark] two shared pairs of electrons between atoms; [1 mark] four electrons remaining on atoms. [1 mark]

Properties of small molecules

1. **a** Covalent bond [1 mark]
   b B [1 mark] or intermolecular forces [1 mark]
2. **a** Its boiling point is below 25 °C. [1 mark]
   b Methane has weak forces between the molecules (intermolecular forces). [1 mark] These weak forces do not require a lot of (heat) energy to break [1 mark], so the boiling point is low.

Polymers

1. **a** One line drawn joining both chains [1 mark]
   b Label points to anywhere along either chain [1 mark]
2. *n* is a large number, to show that the structure in the brackets is repeated many times. [1 mark]

Giant covalent structures

1. **a** X is silicon dioxide/silica; [1 mark] Y is graphite; [1 mark] Z is diamond [1 mark]
   b X/silicon dioxide/silica [1 mark]
2. Any four from: Silicon dioxide has a giant covalent structure. [1 mark] Carbon dioxide is made up of simple molecules. [1 mark] There are strong covalent bonds between the silicon and oxygen atoms in silicon dioxide. [1 mark] There are strong covalent bonds between the carbon and oxygen atoms in carbon dioxide. [1 mark] There are weak intermolecular forces between the carbon dioxide molecules. [1 mark]

Properties of giant covalent structures

1. **a** A pure substance contains only one element or only one compound and so is not a mixture. [1 mark]
   b All of its atoms are bonded together by strong covalent bonds; [1 mark] so a high amount of energy is required to overcome the bonds. [1 mark]
2.

Marks awarded for this type of answer are determined by the Quality of Written Communication (QWC) as well as the standard of your scientific response.
0 marks: No relevant content.
Level 1 (1–2 marks): One or two similarities and differences. The reasons are incorrect or incomplete.
Level 2 (3–4 marks): One or two similarities and differences. They are mostly explained correctly.
Level 3 (5–6 marks): A number of similarities and differences between the physical properties of diamond and graphite have been give. Each has been explained using correct scientific knowledge.
Indicative content Similarities: • Both have a high melting point and a high boiling point. • Both are insoluble in water. • The covalent bonds between the carbon atoms are strong and difficult to break. Differences: • Graphite can conduct electricity, diamond cannot. • In graphite there are (delocalised) electrons between the layers which can move/carry charge. • Diamond has no free electrons. • The overall structure of graphite is weak, the overall structure of diamond is strong. • The intermolecular forces between the layers in graphite are weak so can break with little force.

Graphene and fullerenes

1. Any four from: They both contain just carbon atoms. [1 mark] In both six carbon atoms are arranged in hexagons. [1 mark] They both have delocalised electrons. [1 mark] Graphite consists of many layers. [1 mark] Graphene consists of a single layer. [1 mark]

Nanoparticles

1. a Their diameter is smaller than 100 nm; nanoparticles are between 1 and 100 nm. [1 mark]

   b $\frac{\text{diameter of titanium dioxide particle}}{\text{diameter of a water molecule}} = \frac{10\text{ nm}}{0.1\text{ nm}} = 100$

   [1 mark] $100 = 10^2$, so order of magnitude = 2 [1 mark]

2. a A substance that will increase the rate of a reaction; [1 mark] but is not used up during the reaction. [1 mark]

   b volume = $2 \times 2 \times 2 = 8\text{ cm}^3$; [1 mark] surface area = $(2 \times 2) \times 6 = 24\text{ cm}^2$; [1 mark] ratio is 3:1 (accept 24:8) [1 mark]

   c Smaller pieces increases the surface area [1 mark] which increases the surface area to volume ratio [1 mark]. The larger the surface area to volume ratio, the greater the rate of reaction. [1 mark]

Uses of nanoparticles

1.

Marks awarded for this type of answer are determined by the Quality of Written Communication (QWC) as well as the standard of your scientific response.
0 marks: No relevant content.
Level 1 (1–2 marks): One or two reasons for and against a ban have been stated.
Level 2 (3–4 marks): A number of reasons for and against have been listed.
Level 3 (5–6 marks): A number of reasons for and against a ban have been discussed with valid reasoning.
Indicative content • Nanoparticles have many uses, e.g. medicine, electronics, sun creams, and as catalysts. • There are many potential uses of nanoparticles that will not be discovered if they are banned. • There might not be any risk to human health; more research is needed. • Nanoparticles are only a risk to human health if they are inhaled, like $PM_{2.5}$ particulates in pollution. In the uses stated they cannot be inhaled. • $PM_{2.5}$ particles are larger than the nanoparticles used in these applications, so nanoparticles may be even more dangerous. • Nanoparticles might pose a risk to human health, so a ban would prevent people getting ill. • Because the technology is so new, health problems may become apparent in the future.

Metallic bonding

1. They are arranged in a regular pattern. [1 mark]

   They are attracted by strong bonds. [1 mark]

2. [pointing at circle] positive ion; [1 mark] [pointing at space between circles] sea of electrons/delocalised electrons [1 mark]

3. The electrons are not attached to any particular atom. [1 mark] The electrons are free to move. [1 mark]

4. a Sodium is in Group 1, so has 1 delocalised electron. [1 mark] Calcium is in Group 2 of the periodic table, so has 2 delocalised electrons, which is more than sodium. Calcium also has a higher melting point, so the hypothesis is supported. [1 mark]

   b Find the melting points of more metals from a reliable source, e.g. a data book. [1 mark] Use the periodic table to work out the number of delocalised electrons. [1 mark] Check the link between the number of delocalised electrons and the melting points. [1 mark]

Properties of metals and alloys

1. a A mixture of a metal and another element. [1 mark]

   b A [1 mark]

   c It contains two different types of atoms/metals. [1 mark]

   d The layers of atoms are able to slide over each other; [1 mark] so they can be bent into shapes. [1 mark]

   e An alloy has atoms of different sizes added. [1 mark] These prevent the layers sliding over each over as easily, which means that the alloy is stronger than the pure metal. [1 mark]

Section 3: Quantitative chemistry

Writing formulae

1. b $CuCO_3$ [1 mark]

   c $Mg(NO_3)_2$ [1 mark]

   d $Ca(OH)_2$ [1 mark]

2. a 4 [1 mark] b 8 [1 mark] c 8 [1 mark]

3. a Hydrochloric acid [1 mark]

   b Use an indicator (accept named indicator). [1 mark] Look for a colour change (accept named colour change if indicator was named above e.g. red/orange for universal indicator) [1 mark]

Conservation of mass and balanced chemical equations

1. a $2Mg(s) + O_2(g) \rightarrow 2MgO(s)$

   1 mark for correct symbols/formulae for substances, 1 mark for correct state symbols, 1 mark for correct balancing

   b Mass of magnesium oxide equals mass of the crucible and lid at the end of the heating minus the mass of the crucible and lid at the start. [1 mark]

   c $\frac{1.48 + 1.38 + 1.5}{3} = 1.45$ [1 mark]

   range = $1.5 - 1.38 = 0.12$ [1 mark]

   uncertainty = 1.45 ± 0.06 [1 mark]

   d Some magnesium oxide may have been lost to the air when they opened the crucible. [1 mark] Not all of the magnesium reacted. [1 mark]

e mass of oxygen = mass of magnesium oxide at end of reaction minus mass of magnesium at the beginning of the reaction

$1.45 - 1 = 0.45$ g [1 mark]

Mass changes when a reactant or product is a gas

1. **a** Cotton wool allows the carbon dioxide/gas to escape while preventing any liquid escaping. [1 mark] Using a rubber bung would be a hazard. [1 mark] A rubber bung would prevent the liquid escaping but pressure would build up inside and the bung could suddenly come off with a lot of force/the flask could break/liquid (acid) would likely escape. [1 mark]

   b The mass recorded on the balance would decrease. [1 mark] Carbon dioxide, a gas, is being produced. [1 mark] The carbon dioxide leaves the flask and its mass is no longer recorded by the balance. [1 mark]

   c All of the acid had reacted./The acid was the limiting factor. [1 mark]

   d Uncertainty = the highest result minus the lowest result. [1 mark]

Relative formula mass

1. Add up the relative atomic masses of the atoms. [1 mark]
3. The sulfate ion is SO_4^{2-}, a Group 2 metal forms 2+ ions, so the formula of the compound is XSO_4. [1 mark] M_r of $SO_4 = 32 + (16 \times 4) = 96$ [1 mark] $136 - 96 = 40$; A_r of the metal = 40 [1 mark] (from the periodic table), metal is calcium. [1 mark]
4. The total M_r of reactants = 46 + 60 = 106 [1 mark] The M_r of $C_4H_8O_2 = (4 \times 12) + (8 \times 1) + (2 \times 16) = 88$ [1 mark] So the M_r of Z must be 106 − 88 = 18. [1 mark]

Moles

1. 6.02×10^{23} [1 mark]
2. **b** M_r(NaCl) = 23 + 35.5 = 58.5 g [1 mark]

   mass of 0.5 mol = $0.5 \times 58.5 = 29.25$ g [1 mark]
3. M_r HBr = 1 + 80 = 81 [1 mark]

   $\text{moles} = \frac{\text{mass}}{M_r} = \frac{283.5}{81} = 3.5$ moles [1 mark]
4. $A_r = \frac{\text{mass}}{\text{moles}} = \frac{17.5}{2.5} = 7$ [1 mark]
5. **a** 5 [1 mark] **b** 10

Amounts of substances in equations

1. 3 [1 mark]
2. **a** Put a lit splint to it. [1 mark] They should hear a squeaky 'pop' sound. [1 mark]

   b $\text{number of moles} = \frac{\text{mass}}{\text{relative formula mass}}$;

   mass = number of moles × relative formula mass

   $\text{moles of magnesium} = \frac{1.2}{24} = 0.05$ [1 mark]

   M_r magnesium chloride = $24 + (35.5 \times 2) = 95$ [1 mark]

   mass of magnesium chloride = $0.05 \times 95 = 4.75$ (g) [1 mark]

Using moles to balance equations

1. **a** $\text{number of moles} = \frac{\text{mass}}{\text{relative atomic mass}} = \frac{127}{63.5}$

   = 2 moles [1 mark]
2. **a** hydrogen [1 mark]

   b $\text{moles of Na} = \frac{4.6}{23} = 0.2$ [1 mark]

   $\text{moles of NaCl} = \frac{11.7}{23 + 35.5} = 0.2$ [1 mark]

   ratio = 1:1 [1 mark]

Limiting reactants

1. Magnesium; [1 mark] because it all reacted/there is none left at the end of the reaction. [1 mark]
2. **b** From the equation, the ratio is 1 mole Fe_2O_3 to 2 moles Fe.

   $0.625 \times 2 = 1.25$ moles Fe; [1 mark]

   $1.25 \times 56 = 70$ g [1 mark]
3. $\text{moles Na} = \frac{11.5}{23} = 0.5$ [1 mark] from the equation, the ratio is 2 moles Na to 1 mole Cl_2, $\frac{0.5}{2} = 0.25$ moles of chlorine needed [1 mark] $0.25 \times (35.5 \times 2) = 17.7$ g [1 mark] 17.7 g needed but only 15 g present, means that chlorine is the limiting factor [1 mark]

Concentration of solutions

1. $\frac{100\ cm^3}{1000} = 0.1\ dm^3$ [1 mark]

   mass = volume × concentration = $0.1 \times 15 = 1.5$ g [1 mark]
2. **a** volume in $cm^3 = 0.05\ dm^3 \times 1000 = 50\ cm^3$

   $\text{concentration} = \frac{\text{mass}}{\text{volume}} = \frac{6.2}{50}$ [1 mark]

   $= 0.12\ g/cm^3$ [1 mark]

   b The darker colour shows that the concentration of the copper sulfate has increased. [1 mark] Some water has evaporated, meaning that the volume of the water has decreased. [1 mark] The mass of the copper sulfate has stayed the same because copper sulfate will not evaporate. [1 mark]

3. $\text{concentration} = \frac{\text{mass}}{\text{volume}}$

M_r KI = 19 + 53 = 72 [1 mark]

mass of KI = 0.25 × 72 = 18 g [1 mark]

$\frac{18}{250} = 0.072$ [1 mark] concentration = 0.072 g/cm^3

Using concentrations of solutions in mol/dm³

1. a M_r $CuCl_2$ = 63.5 + (35.5 × 2) = 134.5 [1 mark]

   moles $CuCl_2$ = $\frac{269}{134.5} = 2$ [1 mark]

   concentration = $\frac{2}{2} = 1$ mol/dm^3 [1 mark]

   b $\frac{50}{1000} = 0.05$ dm^3 of solution [1 mark]
   number of moles = 0.05 × 1 = 0.05 mol [1 mark]
   mass = 0.05 × 134.5 = 6.7 g [1 mark]

Amounts of substances in volumes of gases

1. $\text{number of moles} = \frac{\text{mass}}{\text{relative atomic mass}}$

   moles of CO_2 = $\frac{220}{12 + (16 \times 2)} = 5$ [1 mark]

   5 × 24 dm^3 = 120 dm^3 [1 mark]

2. a There would be no more fizzing. [1 mark]

   b mass of CO_2 produced = 256.50 – 256.05 = 0.45 g [1 mark]
   M_r CO_2 = 12 + (16 × 2) = 44 [1 mark]

   moles of CO_2 = $\frac{0.45}{44}$ (= 0.010227…)

   0.010227… × 24 = 0.245 dm^3 [1 mark]

Percentage yield

2. a $\text{number of moles} = \frac{\text{mass}}{\text{relative atomic mass}}$ moles

   $N_2 = \frac{2000}{28} = 71.43$ [1 mark]

   equation shows that moles NH_3 = 2 × moles N_2;
   moles NH_3 = 71.43 × 2 = 142.86 [1 mark]
   M_r NH_3 = 14 + (3 × 1) = 17; mass NH_3 = 142.86 × 17 = 2429 g [1 mark]

   b $\%\ \text{yield} = \frac{\text{mass of product actually made}}{\text{maximum theoretical mass of product}} \times 100$

   $\frac{608}{2429} \times 100 = 25\%$ [2 marks]

   (Allow error carried forward from part a.)

Atom economy

1. a M_r of all reactants = 84 + (2 × 63) = 210 [1 mark]

   M_r of magnesium nitrate = 148

   atom economy = $\frac{148}{210} \times 100$ [1 mark] = 70.5% [1 mark]

   b M_r of all reactants = 24 + 16 + (2 × 63) = 166 [1 mark]

   atom economy = $\frac{148}{166} \times 100$ [1 mark]

   = 89.2% [1 mark]

   c Atom economy for the first reaction is lower. [1 mark] Both reactions produce water but the first reaction produces carbon dioxide also. [1 mark]

   d Any one from: The yield of magnesium nitrate in the reaction is higher; magnesium carbonate is a cheaper reactant than magnesium oxide; they can sell the carbon dioxide produced. [1 mark]

Section 4: Chemical changes

Metal oxides

1. a 4 [1 mark]

   b ethane [1 mark]

   c carbon; [1 mark] hydrogen [1 mark]

2. a iron oxide [1 mark]

   b relative formula mass of iron oxide = (56 × 2) + (16 × 3) = 160

   $\text{moles} = \frac{\text{mass}}{\text{relative formula mass}}$

   moles iron oxide = $\frac{20}{160} = 0.125$ [1 mark]

   from the equation, moles Fe = 0.125 × 2 = 0.25 [1 mark]

   mass Fe = 0.25 × 56 = 14 kg [1 mark]

Reactivity series

1. Lithium loses one electron. [1 mark]
2. Lithium loses an electron less easily than sodium/ Lithium has less tendency to form a positive ion than sodium. [1 mark] This is because lithium's outer electron is nearer to the nucleus (than sodium's). [1 mark] As a result, more energy is needed for lithium's outer electron to be lost (compared to sodium's). [1 mark]
3. a C, A, D, B [1 mark]

   b Any one from: iron, zinc, copper (also accept silver, gold, platinum) [1 mark]

   c Metal B is very reactive; [1 mark] the reaction of metal B with acid could be dangerous/explosive. [1 mark]

Reactivity series – displacement

1. **b** iron; [1 mark] copper [1 mark]
   c $CaSO_4$; [1 mark] Mg [1 mark]
2. The magnesium would go brown/get covered in copper. [1 mark] Copper is being displaced from copper sulfate, [1 mark] because magnesium is more reactive than copper [1 mark]

Extraction of metals

1. **a** Gold is unreactive [1 mark] and so does not form compounds. [1 mark]
   b Zinc is less reactive than carbon. [1 mark] Zinc can be displaced from its compounds by carbon. [1 mark] Aluminium is more reactive than carbon. [1 mark] Aluminium cannot be displaced from its compounds by carbon and electrolysis must be used instead. [1 mark]

Oxidation and reduction in terms of electrons

1. A non-metal ion loses electrons to form atoms. [1 mark]
2. **a** Reduction [1 mark]
   b Oxidation [1 mark]
   c Oxidation [1 mark]
3. **a** $Zn^{2+}(aq) + Mg(s)$ [1 mark] $\rightarrow Mg^{2+}(aq) + Zn(s)$ [1 mark]
   b Reduced: zinc ions [1 mark]
   Oxidised: magnesium atoms [1 mark]

Reactions of acids with metals

1. **a** (s), (aq), (g) [1 mark]
   c The magnesium is oxidised and the hydrogen is reduced. [1 mark]
2. **a** hydrogen; [1 mark] magnesium sulfate [1 mark]
   b Hydrogen ions gain electrons (to form atoms). [1 mark]

Neutralisation of acids and making salts

1. $Mg(NO_3)_2$ [1 mark]
2. **a** A base is insoluble in water while an alkali is soluble in water. [1 mark]
   b sodium chloride; [1 mark] water [1 mark]
3. Add the powder to an acid [1 mark] and collect any gas formed. [1 mark] Add the gas to limewater; [1 mark] if the limewater goes cloudy, the powder is a metal carbonate. [1 mark]

Making soluble salts

1. **a** Metal oxide: zinc oxide [1 mark]
   Acid: sulfuric (acid) [1 mark]
   b Solid metal oxide remained in the beaker (as no more metal oxide will react). [1 mark]
   c Filter the mixture, [1 mark] to remove the excess metal oxide. [1 mark] Heat the solution using a water bath/electric heater and then leave in warm place, [1 mark] to evaporate the water and leave pure crystals. [1 mark]

pH and neutralisation

1. $H^+(aq) + OH^-(aq) \rightarrow H_2O(l)$ [1 mark]
2. (from top, left to right) 7; [1 mark] Weak acid; [1 mark] Purple; [1 mark] 1–2 [1 mark]
3. **a** –; 2; –; 2 [1 mark]
   b Add indicator to the sodium hydroxide. [1 mark] Add acid gradually, mixing; [1 mark] until the indicator just changes colour. [1 mark]

Titrations

1.

Marks awarded for this type of answer are determined by the Quality of Written Communication (QWC) as well as the standard of your scientific response.
0 marks: No relevant content.
Level 1 (1–2 marks): A basic method, which includes some of the equipment.
Level 2 (3–4 marks): There is a clear description of the method, which describes the correct use of equipment. The student has mentioned the need to repeat results.
Level 3 (5–6 marks): A clear, detailed description of the method, which correctly describe how to use the equipment to obtain accurate results.
Indicative content • Use a pipette and safety filler to measure 25 cm³ of sodium hydroxide. • Put the sodium hydroxide into a conical flask with the methyl orange indicator. • Place the conical flask on a white tile. • Fill a burette with hydrochloric acid up to the 0.00 cm³ line. • Gradually add acid to the sodium hydroxide, swirling. • Stop when there is a colour change (yellow to orange). • Read the burette to record the volume of acid. • Repeat until several similar results are obtained (within 0.1 cm³). • Calculate the mean.

2. moles NaOH = volume × concentration = $\frac{25}{1000} \times 0.5$ [1 mark] = 0.0125 [1 mark] from the equation, moles HCl = 0.0125, volume of HCl = $\frac{21.5}{1000} = 0.0215\ dm^3$

   concentration of HCl = $\frac{moles}{volume} = \frac{0.0125}{0.0215}$ [1 mark]

   = 0.58 mol/dm³ [1 mark]

Strong and weak acids

1. Carbonic acid; [1 mark] Ethanoic acid [1 mark]
2. a As the concentration of hydrogen ions decreases the pH increases. [1 mark]
   b 0.000 000 1 = 10^{-7}; [1 mark] pH = 7 [1 mark]
3. a Will have a faster reaction/more vigorous bubbles with the strong acid. [1 mark]
   b Weak acids are only partially ionised in solution; [1 mark] strong acids are completely ionised in solution [1 mark] and so strong acids have a greater concentration of H^+ ions. [1 mark]

The process of electrolysis

1. Copper(II) sulfate solution [1 mark]

   Molten lead bromide [1 mark]
2. a Power supply labelled, [1 mark] electrode on left labelled as 'negative electrode/cathode, [1 mark] electrode on right labelled as 'positive electrode/ anode, [1 mark] liquid labelled as 'electrolyte'. [1 mark]
   b Reduction [1 mark]
3. Copper will be produced at the negative electrode/ cathode. [1 mark] The copper ions are positively charged; [1 mark] opposite charges attract [1 mark] and copper ions gain electrons to form atoms. [1 mark]

Electrolysis of molten ionic compounds

1. The bulb will light. [1 mark] The ions in the molten zinc(II) chloride are free to move and so conduct electricity. [1 mark] Bubbles of gas (chlorine) form at the positive electrode/anode; [1 mark] as (negative) chloride ions are attracted to the positive electrode and lose electrons to form chlorine atoms which bond to form chlorine molecules. [1 mark] Silvery coloured metal forms at the negative electrode/cathode [1 mark] as (positive) zinc ions are attracted to the negative electrode and gain electrons to form zinc atoms. [1 mark]
2. $2Br^- - 2e^-$ [1 mark] $\rightarrow Br_2$ [1mark]
   or $2Br^-$ [1 mark]
   $\rightarrow Br_2 + 2e^-$ [1 mark]

Using electrolysis to extract metals

1. Potassium [1 mark]

   Sodium [1 mark]
2. a Aluminium is more reactive than carbon. [1 mark]
   b The mixture has a lower melting point than pure aluminium oxide; [1 mark] as a result, less energy is needed. [1 mark] This reduces the cost of the process. [1 mark]
   c Aluminium is discharged at the cathode. The cathode has a negative charge; [1 mark] and aluminium ions have a positive charge. [1 mark]

   Oxygen is discharged at the anodes. The anodes have a positive charge; [1 mark] and oxide ions have a negative charge. [1 mark]

Electrolysis of aqueous solutions

1. Cathode: hydrogen; [1 mark] sodium [1 mark]
2. a To prevent a short circuit [1 mark]
   b Sodium is more reactive than hydrogen and it is the less reactive element that is discharged. [1 mark]
   c Bleaching of blue litmus is a test for chlorine gas, so chlorine gas is being produced. [1 mark]
   d Measure the volume of one gas produced [1 mark] at a range of different voltages. [1 mark] One from: Control variable: concentration of sodium chloride solution/volume of sodium chloride solution/time over which gas is collected [1 mark]

Half equations at electrodes

1. It happens at the anode. [1 mark]

   It shows oxidation. [1 mark]
2. a O_2; [1 mark] $4e^-$ [1 mark]
   b Cu^{2+} [1 mark]+ $2e^-$ [1 mark] $\rightarrow Cu$
   c As the current increases, the mass of copper deposited will increase. [1 mark] This happens because there are more electrons travelling around the circuit (in a fixed amount of time). [1 mark]
   d Repeat the experiment. [1 mark] Check that the mass of copper deposited is similar in each repeat. [1 mark]

Section 5: Energy changes in reactions

Exothermic and endothermic reactions

1.

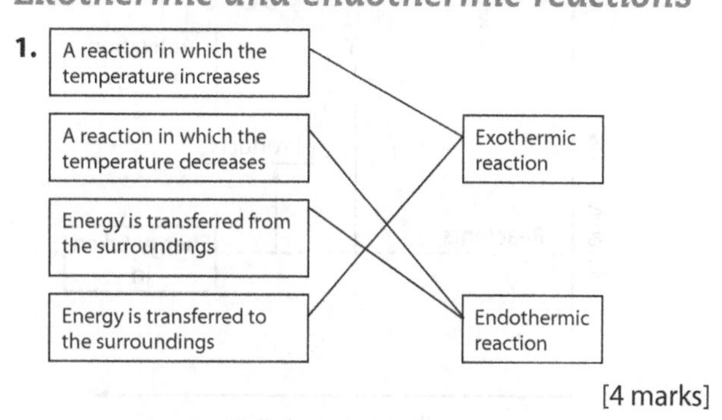

[4 marks]

2. a −17 °C [1 mark]
   b Endothermic [1 mark]

3 a

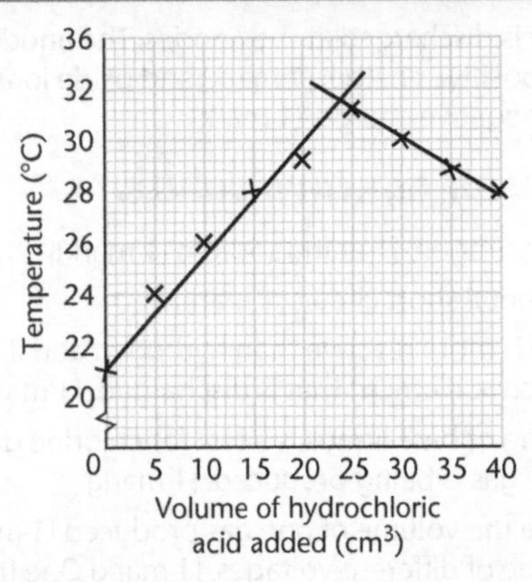

1 mark for each line

b The temperature increases to a maximum [1 mark] and then starts to drop slowly. [1 mark]

c 31.5 °C [1 mark]

d Exothermic; [1 mark] because the temperature of the surrounding increases. [1 mark]

e The chemical reaction is over. [1 mark] No more heat energy is released so as more acid is added it just cools the solution. [1 mark]

Reaction profiles

1. a, b

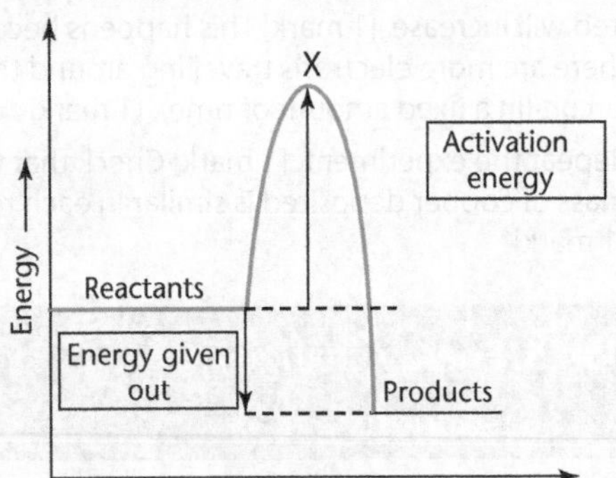

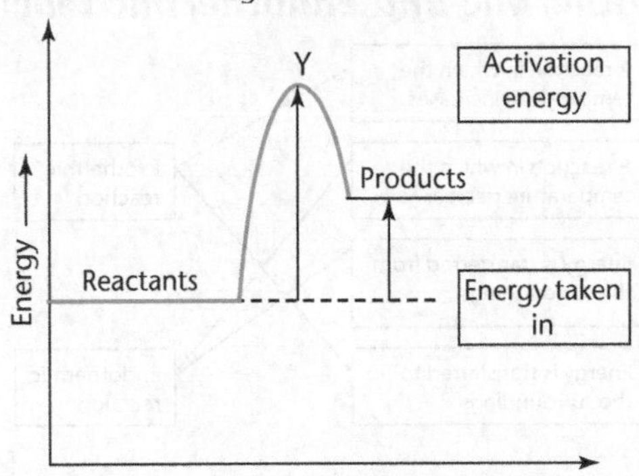

2 marks for completing the labels for reaction X and 2 marks for completing the labels for reaction Y.

c Profile Y; [1 mark] because the relative energy of the products is greater than that of the reactants; [1 mark] indicating that energy has been taken in from the surroundings. [1 mark]

2. a

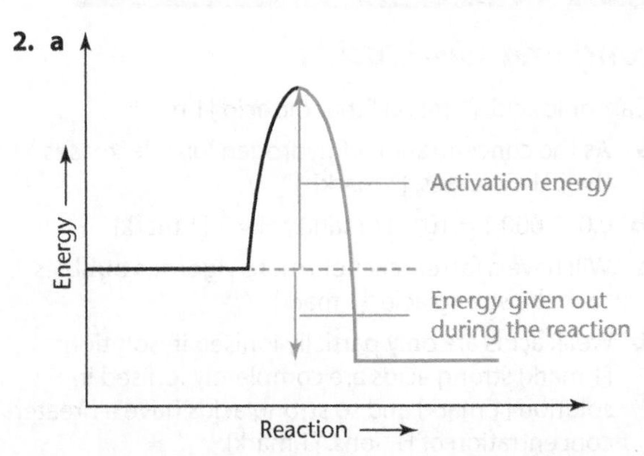

Correctly completed profile [1 mark]; correct arrow for energy change [1 mark]; activation energy label. [1 mark]

b The activation energy is the minimum energy [1 mark] required to start a reaction. [1 mark]

Energy change of reactions

1. a The H–H bonds are broken; [1 mark] the Cl–Cl bonds are broken; [1 mark] new H–Cl bonds are made. [1 mark]

c 2HBr(g) [1 mark]

d Reactants: H–H = 436; Br–Br = 193; [1 mark] total bond energies of reactants = 629 kJ/mol. [1 mark] Products: 2 × H–Br = 2 × 336; [1 mark] total bond energies of products = 672 kJ/mol. [1 mark] Overall energy change = difference between the reactants and products, 629 – 672 = –43 kJ/mol. [1 mark] Therefore, the prediction is correct because –43 is less negative than –179, meaning that less energy is given out during the reaction. [1 mark]

Cells and batteries

1. batteries; [1 mark] cells; [1 mark] series; [1 mark] voltage [1 mark]

2. a When one electrode is lifted out the voltage will decrease from 0.8 V to 0.0 V. [1 mark]

b 2; [1 mark] 2 [1 mark]

c It is to complete the circuit/allow the electrons to flow between the two beakers. [1 mark]

d The voltage of the cell will increase. [1 mark]

Fuel cells

1. a $2H_2(g) + O_2(g) \rightarrow 2H_2O(l)$

1 mark for correct formulae; 1 mark for balancing; 1 mark for correct state symbols

b $2H_2 \rightarrow 4H^+ + 4e^-$ [1 mark]

$4H^+ + O_2 + 4e^- \rightarrow 2H_2$ [1 mark]

c No pollutants are formed at point of use as only product is water. [1 mark] It can be a renewable source of power if the energy used to make the hydrogen comes from a renewable source. [1 mark]

d Hydrogen is explosive. [1 mark] Hydrogen is difficult to store. [1 mark]

Section 6: Rates of reactions

Measuring rates of reaction

1. Reaction between marble chips and hydrochloric acid [1 mark]
2. **a** magnesium + hydrochloric acid → magnesium chloride + hydrogen [1 mark]
   b The measuring cylinder should be filled with water. [1 mark] There should be a bung in the top of the conical flask. [1 mark]
   c Most of the hydrogen gas would escape through the top of the conical flask. [1 mark] It would not be possible to measure the amount of gas in the measuring cylinder because it would not be visible. [1 mark]
   d At the start [1 mark]
   e 65 cm^3 [1 mark] **f** 37 s [1 mark]

Calculating rates of reaction

1. **a** The change in mass of the reactant vessel [1 mark]
   b mean rate of reaction = $\frac{\text{mass or volume of gas produced}}{\text{time taken}}$ [1 mark]
   c $\frac{\text{volume of gas produced}}{\text{time taken}} = \frac{20\ cm^3}{40\ s} = 0.5\ cm^3/s$
   1 mark for correct answer and 1 mark for correct units
2. **a**

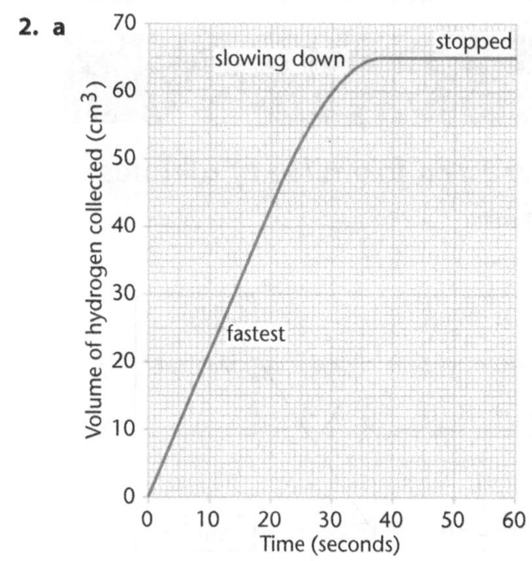

1 mark per answer

Effect of concentration and pressure

1. chemical, [1 mark] particles, [1 mark] sufficient, [1 mark] activation, [1 mark] concentration, [1 mark] increases [1 mark]
2. The rate of reaction depends on the concentration of the acid. [1 mark]
3. **a** Mix the sodium thiosulfate and hydrochloric acid in a beaker. [1 mark] Put the beaker on a black cross. [1 mark] Time how long the cross will take to disappear. [1 mark]
4. The volume will increase and the particles will spread out to fill the space. This decreases the pressure. [1 mark] Particles will collide less frequently [1 mark] therefore the chance of successful collisions will decrease and so will the rate of reaction. [1 mark]

Effect of surface area

1. **a** A [1 mark] because there is only one piece of calcium carbonate. [1 mark]
   b B [1 mark] because there are more calcium carbonate particles available with more surface area for the acid to react with. [1 mark]
2. **a** To make it a fair test [1 mark]
   b Surface area [1 mark]
   c If you increase the surface area of a reactant, you increase the rate of reaction. [1 mark]
   d Repeat each experiment. [1 mark] Carry out further experiments with different size pieces of rhubarb. [1 mark]

Effect of temperature

1. **a**

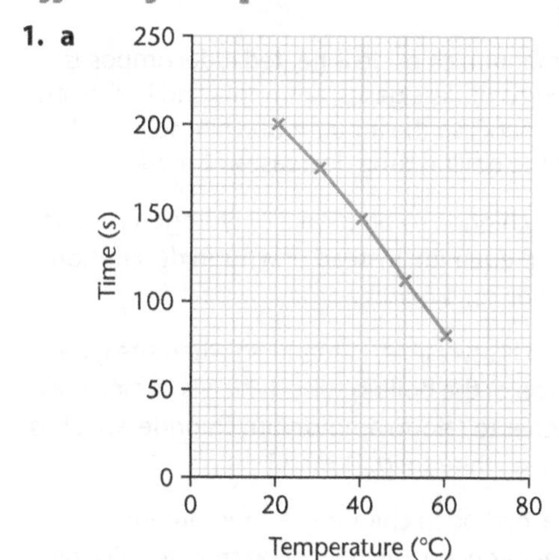

   Scales, [1 mark] labels, [1 mark] points plotted correctly, [1 mark] line of best fit [1 mark]
   b As the temperature increases the time taken for the reaction decreases, meaning that the rate of reaction increases. [1 mark] This is because the reaction particles have more energy [1 mark] so they move quickly. The particles collide more often [1 mark] and more of the collisions are successful [1 mark] so the rate increases.
   c The student should repeat each experiment and then take the mean value. [1 mark] Use the mean value to plot the graph. [1 mark]

Effect of a catalyst

1. Catalysts are specific to one single reaction. [1 mark]
   Catalysts are substances that change the speed of a chemical reaction. [1 mark]

2. a A new 'hump' line drawn below the one already on the profile. The activation energy is lower. [1 mark]
   b hydrogen peroxide → water + oxygen [1 mark]
   c Measure the time [1 mark] it takes to collect 40 cm^3 (or fixed amount) of oxygen gas [1 mark] in a gas syringe (or by displacement with water) and repeat for different catalysts. [1 mark]
   d To act as a control – so there is something to compare the other results with. [1 mark]
   e Liver contains enzymes, which are biological catalysts. [1 mark]
   f Manganese oxide [1 mark] because it produced the required amount of gas the quickest. [1 mark]

Reversible reactions and energy changes

1. Reversible [1 mark]
2. a This is a reversible reaction. [1 mark]
   b It must be heated. [1 mark] The forward reaction is endothermic [1 mark] so heat energy must be transferred into the system. [1 mark]
   Alternative answer: Add a dehydrating agent, e.g. conc. sulfuric acid; [1 mark] which removes the water; [1 mark] pulling the equilibrium to the right. [1 mark]
   c Add water to the anhydrous copper sulfate. [1 mark]
3. When ammonium chloride is heated it decomposes to form two colourless gases, ammonia and hydrogen chloride [1 mark]; so the amount of white solid at the bottom of the boiling tube decreases. [1 mark]

   ammonium chloride → ammonia + hydrogen chloride (accept word equation instead of written description of the reaction for 1 mark)

   The reaction is reversible. [1 mark] So when the gases reach the top of the boiling tube, which is cooler, they react together to reform ammonium chloride, which is observed as a white solid. [1 mark]

   ammonia + hydrogen chloride → ammonium chloride (accept word equation instead of written description of the reaction for 1 mark)

Equilibrium and Le Chatelier's Principle

1. The rates of the forward and reverse reactions are the same. [1 mark]
2. To the right [1 mark]
3. The fizzing noise is CO_2 gas escaping from the bottle. [1 mark] When the bottle is open the equilibrium is lost [1 mark] and the gas escapes as soon as it is formed. [1 mark]
4. If a system is at equilibrium and a change is made to any of the conditions, then the system responds to counteract the change. [1 mark]
5. a The reaction mixture will go from brown to colourless. [1 mark]
   b The reaction mixture will go from colourless to brown. [1 mark]

Changing the position of equilibrium

1. a $3H_2(g)$; [1 mark] $2NH_3(g)$ [1 mark]
   b Increase the pressure; [1 mark] lower the temperature. [1 mark]
2. a $\rightleftharpoons C_2H_5OH(g)$
   equilibrium sign, [1 mark] the correct state symbol, [1 mark] formula of ethanol [1 mark]
   b The position of equilibrium will move to the left. [1 mark] If the reaction to the right is exothermic then the reaction to the left must be endothermic and is favoured by heating a reaction. [1 mark]
   c The position of equilibrium would move to the right [1 mark] as two molecules of gas would rearrange into one molecule. [1 mark] Increasing the pressure favours the side of the reaction with fewer molecules of gas.
   d Catalysts have no effect on the position of equilibrium. [1 mark]
   e The forward reaction favours low temperatures [1 mark] but if the temperature is too low the rate of reaction will be very slow [1 mark] and it becomes economically unviable.
   The forward reaction favours high pressures [1 mark] but high pressures require very expensive equipment. [1 mark] So in practice the industry tries to get a good balance between rate, cost and yield.

Section 7: Organic chemistry

Structure and formulae of alkanes

1. C_3H_8 [1 mark]
2. C_2H_6 [1 mark]
3.

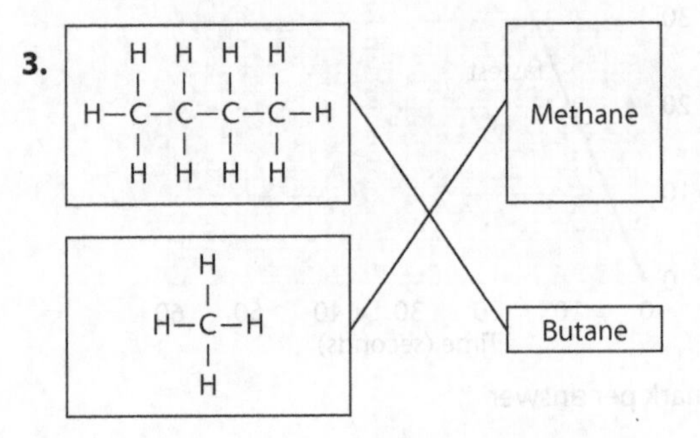

4. a C_6H_{14} [1 mark] b $C_{25}H_{52}$ [1 mark]

Fractional distillation and petrochemicals

1. mixture, [1 mark] distillation, [1 mark] evaporated, [1 mark] boiling [1 mark]
2. Fuel oil is a much bigger molecule than paraffin and so has a higher boiling point than paraffin. [1 mark] All fractions boil and the vapours rise up the column. [1 mark] Those with higher boiling points condense first and so are collected near the bottom of the column. [1 mark]
3. **a** A barrel from well A contains $\frac{15}{100} \times 150 = 22.5$ kg.

   A barrel from well B contains $\frac{20}{100} \times 142 = 28.4$ kg.

   The barrel from well B contains the most paraffin.

   1 mark for working out the percentages, 1 mark for working out which barrel contains the most paraffin.

Properties of hydrocarbons

1. Large hydrocarbon molecules are more viscous than small hydrocarbon molecules. [1 mark]

   Methane is easier to ignite than butane. [1 mark]
2. **b** 23 °C [1 mark]

   c As the number of carbon atoms increases, the boiling point also increases. [1 mark]

   d As alkane molecules get bigger, the forces between the molecules also get bigger. [1 mark] So more energy is needed to enable the molecules to go from the liquid state to the gaseous state. [1 mark]

Combustion of fuels

1. carbon, carbon monoxide [1 mark] (both need to be correct)

Cracking and alkenes

1.

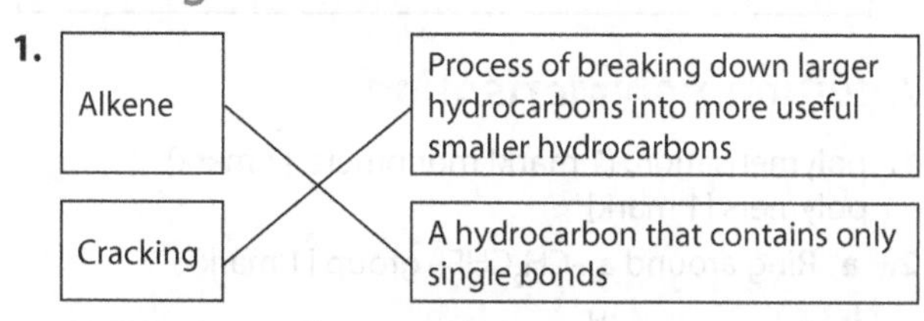

2. **a** Porcelain chips and the mineral wool need to be swapped around. [1 mark] End of delivery tube should be under the test tube to collect the gas. [1 mark]

   b 3 [1 mark]

   c C_4H_8 and C_2H_4 [1 mark]

   d Add a few drops of bromine water to the gaseous product. [1 mark] If the bromine water decolourises then alkenes are present [1 mark] and the reaction has been successful. If it remains orange-brown then only alkanes are present [1 mark] and the reaction has not been successful.

Structure and formulae of alkenes

1.

Ethene	2	C_2H_4	$CH_2=CH_2$
Propene	3	C_3H_6	$CH_3-CH=CH_2$
Butene	4	C_4H_8	$CH_3-CH=CH-CH_3$
Pentene	5	C_5H_{10}	$CH_3-CH=CH-CH_2-CH_3$

[1 mark for each correct row]

2. **a** C_6H_{12} [1 mark] **b** C_8H_{16} [1 mark]
3. Propene contains a carbon–carbon double bond which makes it more reactive than propane which only has carbon–carbon single bonds. [1 mark] When bromine water is added to propene, it decolourises [1 mark] but when it is added to propane it remains brown. [1 mark]

Reactions of alkenes

1. more, [1 mark] double, [1 mark] addition, [1 mark] saturated [1 mark]
2. **a**

```
   H  H
   |  |
H—C—C—H
   |  |
   H  H
```

   b ethane [1 mark]

   c The hydrogen has been added into the molecule across the double bond. [1 mark]
3.

```
   Cl Cl H
   |  |  |
H—C—C—C—H
   |  |  |
   H  H  H
```

   [1 mark]
4. **a** Alkanes burn with a clean flame and alkenes burn with a smoky flame. [1 mark]

   b Alkanes undergo complete combustion but alkenes undergo incomplete combustion. [1 mark]

   c Any of the alkanes in the table. [1 mark] Don't want the cooking pans covered in soot. [1 mark] / Don't want to make carbon monoxide (or any other reasonable answer).

Structure and formulae of alcohols

1. **a**

```
   H  H  H
   |  |  |
H—C—C—C—O—H
   |  |  |
   H  H  H
```

   [1 mark]

   b Propanol [1 mark]

2. Formula of ethane is C_2H_6; general formula of alkanes is C_nH_{2n+2}. [1 mark]

   Formula of ethanol is C_2H_5OH; general formula of alcohols is $C_nH_{2n+1}OH$. [1 mark]

   Functional group of alkanes only contains C–H single bonds. Also referred to as saturated, as they are unreactive. [1 mark]

   Functional groups of alcohols is –OH. The –OH group makes them more reactive and more soluble in water. [1 mark]

   Accept any correct alkane and alcohol as an example.

3. a Conical flask: bubbles [1 mark]
   Boiling tube: limewater goes cloudy [1 mark]
   b Carbon dioxide [1 mark]
   c Yeast/enzyme [1 mark]
   d If too hot the enzymes will be denatured/destroyed and won't work; [1 mark] if too cold the enzymes will be inactive. [1 mark]
   e glucose → ethanol + carbon dioxide [1 mark]

Uses of alcohols

1. Many organic substances are soluble in it. [1 mark]
2. a Feedstock is the main raw material used in an industrial process to manufacture a product. [1 mark]
   b Alcoholic drinks [1 mark]
3. a ethanol + oxygen → carbon dioxide + water [1 mark]
   b $3O_2$; [1 mark] $2CO_2$ [1 mark]
4. Mix an alcohol and carboxylic acid [1 mark] in the presence of a concentrated acid catalyst. [1 mark]

   alcohol + carboxylic acid → ester + water

   Example is ethyl ethanoate, formed from ethanol and ethanoic acid. [1 mark]

Carboxylic acids

1. 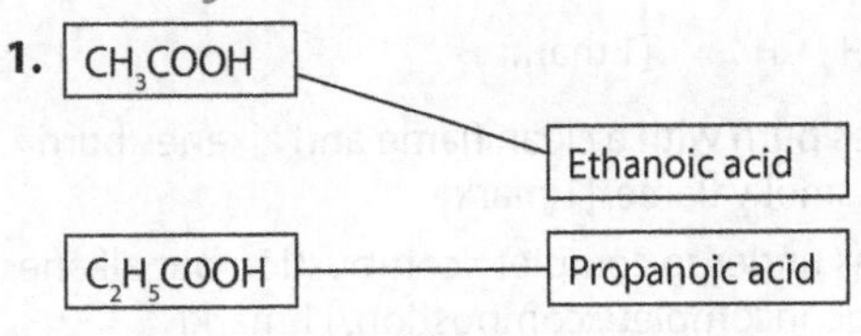

2. potassium carbonate; [1 mark] water; [1 mark] carbon dioxide [1 mark]
3. The H^+ ion is the ion that makes solutions acidic. [1 mark] In strong acids all the particles undergo complete ionisation when dissolved in water but in weak acids the ionisation is not complete. [1 mark] When comparing acids of the same concentration, hydrochloric is a strong acid because only H^+ and Cl^- ions exist in solution [1 mark] and the H^+ ion concentration is high. Ethanoic acid is a weak acid because ethanoic acid molecules and H^+ and CH_3COO^- ions all exist in solution [1 mark] and the H^+ ion concentration is low.
4.

Marks awarded for this type of answer are determined by the Quality of Written Communication (QWC) as well as the standard of your scientific response.
0 marks: No relevant content.
Level 1 (1–2 marks): There is a brief comparison with limited examples
Level 2 (3–4 marks): There is some comparison with examples
Level 3 (5–6 marks): There is a detailed comparison with clear examples
Indicative content • Formula of ethanoic acid is CH_3COOH. • Formula of ethanol is C_2H_5OH; general formula is $C_nH_{2n+1}OH$. • Functional group of carboxylic acid is –COOH. • Functional group of alcohols is –OH. • Both smaller alcohols and carboxylic acids are soluble in water and the rest are partially soluble. • Both carboxylic acids and alcohols burn in oxygen to produce carbon dioxide and water. In both cases incomplete combustion occurs if the oxygen supply is limited. • Carboxylic acids are weak acids, pH 4–6; alcohols are considered to be around neutral (very slightly basic approx. 7.3 – but accept neutral). • Carboxylic acids are reactive – they will react with sodium metal to produce a salt and hydrogen gas; also react with metal carbonates and metal hydroxides. This is due to the presence of the –COOH group. Alcohols will react slowly with Na metal but not the other reactants. • Carboxylic acids and alcohols will react together to produce an ester and water.

Addition polymerisation

1. polymerisation, [1 mark] monomers, [1 mark] polymers [1 mark]
2. a Ring around a $–CH_2CHF–$ group [1 mark]
   b 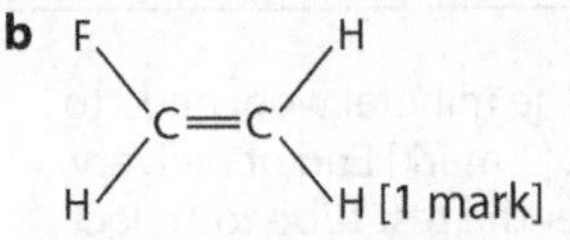

3. a A very large number/many monomers [1 mark]
   b To show the repeating unit [1 mark]
   d During the reaction the carbon–carbon double bonds in the ethene molecules open up [1 mark] and are replaced by single carbon–carbon bonds [1 mark] as the monomers join together. [1 mark]
   e Poly(butene) [1 mark]

Condensation polymerisation

1. **a** Condensation polymerisation [1 mark]
   b A small molecule, such as water [1 mark]
   c HO–CH_2CH_2–OH [1 mark]
2. **a** Both –OH groups should be ringed; [1 mark] both –COOH groups should be ringed. [1 mark]
   b Highlight parts inside the brackets. [1 mark]
   c Because a small molecule, here it is water, has been lost during the reaction. [1 mark]
3. Each process involves the joining of small monomer molecules to make a large polymer molecule [1 mark]. In the case of poly(ethene), there is only one type of monomer molecule. The double bond opens up during the reaction [1 mark] and all the monomers join together. No other compound is formed. [1 mark] Nylon is made from two different monomers with different functional groups [1 mark]. During the reaction a small molecule such as water is also formed. [1 mark]

Amino acids

1. Condensation [1 mark]
2. **a** Ring the –NH_2 group [1 mark] and the –CO_2H group. [1 mark]
   b water [1 mark]
   c

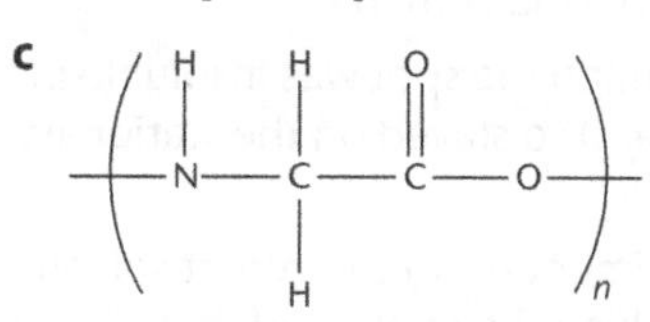

   or (–$HNCH_2COO$–)$_n$ [1 mark]

3.

Marks awarded for this type of answer are determined by the Quality of Written Communication
(QWC) as well as the standard of your scientific response.
0 marks: No relevant content.
Level 1 (1–2 marks): There is a brief description of the process.
Level 2 (3–4 marks): There is an explanation using correct scientific terms.
Level 3 (5–6 marks): There is a detailed, explanation with examples.

Indicative content

- Amino acids are the building blocks of polypeptides.
- Humans use 20 different amino acids.
- Polypeptides are the building blocks of proteins.
- Amino acids are small molecules with –NH_2 and –CO_2H functional groups (but note that some also have other functional groups).
- Amino acids consist of a central carbon atom linked to an amino group (–NH_2), an acid group (–CO_2H), a hydrogen atom, and a variable side chain, e.g. glycine or alanine Amino acids react by condensation polymerisation.
- A long chain molecule is formed and water molecules are also formed during the reaction.
- A single amino acid such as glycine can combine to produce poly(glycine).
  $nH_2NCH_2COOH \rightarrow (-HNCH_2COO-)_n + nH_2O$
- Different amino acids can join to produce a whole range of different proteins – accept any correct equation.
- DNA provides the information about the order of amino acids in proteins.

DNA and other naturally occurring polymers

1. Glucose [1 mark]
2. **a** —OH—□—OH—□—OH—□—
   1 mark for correct repeating pattern, 1 mark for leaving unconnected bonds at the ends
   b Water [1 mark]
3. **a** DNA [1 mark]
   b It encodes genetic instructions for the development and functioning of living things. [1 mark]
4. **a** From top to bottom: phosphate; [1 mark] sugar; [1 mark] base [1 mark]
   b Circle the nucleotide (to include the sugar and phosphate group) with 'A'. [1 mark]

Section 8: Chemical analysis

Pure substances, mixtures and formulations

1. **a** Box A [1 mark]
   Reason: It shows a compound made from two different atoms. [1 mark]
   b Box D [1 mark]
   Reason: It shows two different elements. One element is single atoms (He). [1 mark] The other element is diatomic (H_2). [1 mark]
2. To make sure the hand cream does the job it is meant to. [1 mark] To prevent it from harming your skin. [1 mark]
3. A pure substance only contains one type of particle. [1 mark] This product is a mixture of vitamin C, calcium, fibre, water, thiamine, etc. [1 mark] Therefore it is not pure. [1 mark]
4. Add water to the mixture and stir so that the salt dissolves. [1 mark] Pour the mixture through a filter funnel. As the sand is insoluble it will stay on the filter paper. [1 mark] Transfer the filtrate to an evaporating basin and gently heat to evaporate off the water. [1 mark] Leave the salt to crystallise out as the solution cools. [1 mark]

5. The melting point taken from the data book represents a pure sample [1 mark] because it has a sharp melting point. [1 mark] Impure samples have lower melting points and they melt over a wider temperature range. [1 mark] Therefore the student's sample was not pure. [1 mark]

Chromatography and R_f values

1. a Sample A [1 mark]
   b As the solvent/mobile phase moves up through the paper, it carries the dissolved substances with it. [1 mark] Each substance travels a different distance [1 mark] due to the different forces of attraction between each substance and the chromatography paper/stationary phase or the solvent. [1 mark]
2. a

Marks awarded for this type of answer are determined by the Quality of Written Communication (QWC) as well as the standard of your scientific response.
0 marks: No relevant content.
Level 1 (1–2 marks): There is a brief description of the method.
Level 2 (3–4 marks): There is a description of the method using correct scientific terms.
Level 3 (5–6 marks): There is a detailed description of the method, giving steps in the correct sequence and using correct scientific terms.
Indicative content • Draw a horizontal pencil line about 2 cm from the bottom of the chromatography paper. • Use a capillary tube to put a spot of each sample on the pencil line. • Label the spots in pencil. • Pour the water/solvent into a beaker/ chromatography tank to a depth of approximately 1 cm. • Clip the top of the chromatography paper to wooden spill/glass rod. • Hang/suspend the chromatography paper into the beaker/chromatography tank by resting the spill/glass rod on the top edge of the container. • Make sure the bottom of the chromatography paper just dips into the water/solvent. The pencil line and spots of sample must be above the solvent. • Leave the beaker/chromatography tank until the solvent has travelled about three quarters way up the chromatography paper. • Carefully remove the paper. • In pencil mark the solvent front and hang the paper up to dry.

   b The sample has three components, but only two can be identified. [1 mark] The pale blue spot is E133. [1 mark] The yellow spot is E102. [1 mark]
   c Work out the R_f value of the green spot and then look it up in a data book to find out what it is [1 mark] *or* Repeat the experiment using different known green dyes. [1 mark]
   d $R_f = \dfrac{\text{distance spot moved}}{\text{distance solvent moved}}$; [1 mark]

   $R_f = \dfrac{10.2}{14.5} = 0.70$ (to 2 sig figs) [1 mark]
3. a Distance moved by solvent $= \dfrac{\text{distance moved by spot}}{R_f}$

   [1 mark] $= \dfrac{12.0}{0.86} = 14.0$ [1 mark]
   b Distance travelled by B $= R_f \times$ distance travelled by solvent [1 mark] $= 0.41 \times 14.0 = 5.74$ cm $= 5.7$ cm to 2 significant figures (units and significant figures correct for the 2nd mark) [1 mark]
4. a The mixture contains two colours. [1 mark] They are blue and red. [1 mark]
   b To increase the reliability of the results. [1 mark]
   c One from: they may have drawn/measured the position of the solvent front incorrectly; measured the distance to the of the spot incorrectly; accidently used a different solvent. [1 mark]
   d The chemical forming the spot was insoluble in the solvent [1 mark] and so stayed on the stationary phase. [1 mark]
   e Repeat the experiment using a different solvent; [1 mark] one in which the 'spot' is soluble. [1 mark] Calculate the R_f value and then use a data book. [1 mark]

Tests for common gases

1. Put a glowing splint in the gas and if it relights the gas present is oxygen. [1 mark] Add some limewater to the test tube, shake it and if it goes cloudy/milky the gas present is carbon dioxide. [1 mark]
2. a No water in the test tube – needs to collect over water. [1 mark] Delivery tube needs a bung otherwise the gas will escape. [1 mark]

Flame tests

1. a $CuCO_3$ [1 mark]
   b $CuCO_3(s) + 2HCl(aq) \rightarrow CuCl_2(aq) + H_2O(l) + CO_2(g)$
   1 mark for correct formulae, 1 mark for balancing the equation, 1 mark for correct state symbols
2. a Dip a clean nichrome wire into the metal ion solution. [1 mark] Hold the tip of the wire in a blue Bunsen burner flame. [1 mark]
   b Yellow [1 mark]
   c The result would still be a yellow flame. [1 mark] The sodium ions in the sample would mask the lilac colour from the potassium ions. [1 mark]

Metal hydroxides

1. A precipitation reaction is when two solutions react together; [1 mark] producing a solid and another solution. [1 mark]
3. a A light-green precipitate forms. [1 mark]
   b $FeSO_4(aq) + 2NaOH(aq) \rightarrow Fe(OH)_2(s) + Na_2SO_4(aq)$
   1 mark for correct reactants and products, 1 mark for correct balancing, 1 mark for correct state symbols
4. a copper(II) [1 mark] b $2OH^-$ [1 mark]

Tests for anions

1. a [in order down table] Sulfate ions not present; [1 mark] Iodide ions present; [1 mark] Aluminium, calcium or magnesium ions present; [1 mark] Calcium or magnesium ions present; [1 mark] Calcium ions present; [1 mark]
   b Calcium iodide [1 mark]
   c Add silver nitrate solution and dilute nitric acid to a solution of the compound; [1 mark] the expected result is a creamy precipitate. [1 mark] Add sodium hydroxide solution; [1 mark] the expected result is a green precipitate. [1 mark]

Instrumental methods

1. Instrumental methods are more accurate, [1 mark] more sensitive [1 mark] and faster [1 mark] than chemical tests.
2. a 6 [1 mark]
   b Methane, carbon dioxide [1 mark]
   c Hydrogen, [1 mark] because it has the smallest peak [1 mark]
3. [The sample is put into a flame and] the light given out is passed through a spectroscope. [1 mark]
   [The output is a line spectrum that can be analysed to] identify metal ions in solution and measure their concentrations. [1 mark]

Section 9: Chemistry of the atmosphere

The Earth's atmosphere – now and in the past

1. a Nitrogen [1 mark]
   b One from: carbon dioxide, water vapour, any noble gas (not hydrogen) [1 mark]
2. a copper + oxygen → copper oxide [1 mark]
   b $\frac{11.5}{60.0} \times 100 = 19.2\%$
   1 mark for working out the amount of oxygen ($60\ cm^3 - 48.5\ cm^3$); 1 mark for converting it to a percentage.
   c Two from: Reaction not carried out for long enough for all the oxygen in the apparatus to react. [1 mark] The copper was not in excess. [1 mark] An oxide layer stops copper underneath from reacting even if the copper is in excess. [1 mark]
3. a One from: The evidence is limited. [1 mark] Scientists must rely on models built on assumptions. [1 mark] New evidence from improvements in technology causes scientists to re-think previous theories. [1 mark]
   b Assumption: Volcanoes give out a similar composition of gases today as they did billions of years ago. [1 mark] Evidence: The composition of gases vented out of volcanoes today. [1 mark]
   Assumption: The Earth's atmosphere may have been like Mars or Venus today and there was lots of carbon dioxide. [1 mark] Evidence: Scientists study the atmospheres of different planets. [1 mark]
   Assumption: As the oceans formed carbon dioxide dissolved in the water and carbonates precipitated out. [1 mark] Evidence: Carbon and boron isotope ratios can be measured in sediments under the sea and linked directly back in time. [1 mark]
   Assumption: The number of stomata on leaves can be linked to the levels of carbon dioxide in the early atmosphere. [1 mark] Evidence: The number of stomata found on fossilised leaves. [1 mark]
   (Accept other reasonable linked assumptions and evidence.)

Changes in oxygen and carbon dioxide

1. a 6; 6; 6 (1 mark for balancing reactants and 1 mark for balancing the products)
   b Carbon dioxide levels decreased as plants used it up. [1 mark] Oxygen levels increased as plants made it. [1 mark]
   c At first there was no oxygen [1 mark] because as the oxygen was produced by the algae it reacted with the iron to form iron oxide. [1 mark] When all the iron was used up the iron oxide reaction stopped. [1 mark] As more oxygen was made it went in to the atmosphere and the level started to rise. [1 mark]

Greenhouse gases

1. a Carbon dioxide; [1 mark] methane [1 mark]
   b Short wavelength radiation from the Sun enters the atmosphere. [1 mark] Some of the radiation transfers into thermal energy when it reaches the Earth, while some of it is reflected back to space. [1 mark]
   Long wavelength radiation radiated back from Earth is absorbed by greenhouse gases. [1 mark] A gradual build-up of greenhouse gases means that less of the heat is radiated back to space, causing an increase in temperature. [1 mark]
2. a Parts per million [1 mark]
   b From the graph, the concentrations were 370 ppm in 2000 and 399 ppm in 2014. Allow readings of

+/– 1 ppm, which will then need to be carried through the calculation.

The increase is 399 – 370 = 29 ppm [1 mark]

$\% \text{ increase} = \frac{29}{370} \times 100 = 7.83784\%$ [1 mark]

= 7.84% to 3 significant figures [1 mark]

c CO_2 is a greenhouse gas, so as levels increase so will the average global temperatures, which may lead to global warming. [1 mark] You need to look at the levels of CO_2 over a much longer period of time, e.g. several 100 000 years, before you can reach a meaningful conclusion. [1 mark]

d Increase in the combustion of fossil fuels. [1 mark] Increase in deforestation. [1 mark]

Global climate change

1. Increase in agriculture [1 mark]
2. Glaciers are mountain reserves of fresh water that are frozen. [1 mark] As the climate warms, more of the ice starts to melt and the glaciers retreat. [1 mark] Scientist can measure the rate at which the glaciers retreat, [1 mark] providing first-hand data.
3.

Marks awarded for this type of answer are determined by the Quality of Written Communication (QWC) as well as the standard of your scientific response.
0 marks: No relevant content.
Level 1 (1–2 marks): There is a brief description of an environmental change and possible impacts.
Level 2 (3–4 marks): There is description of an environmental change linked to several possible consequences and impacts.
Level 3 (5–6 marks): There is a detailed description of environmental changes with possible consequences and impacts.
Examples of the chemistry points made in the response:
Increased temperatures: temperature stress for humans and wildlife • too hot to make a living • some species may eventually die out • adapt buildings so they will cope with hotter temperatures • changes to food production capacity: see different crops growing in different parts of the world due to changes in temperature and annual rainfall • farmers in some parts of the world may not be able to farm any more; changes could affect the food chain. More extreme weather events, such as wildfires, droughts, flooding, tornadoes, etc. • no fresh water supplies in some areas, which will lead to mass migration to different countries – could be as a result of drought or flooding • increase in the spread of disease due to poor sanitation as a result of flooding/drought • species becoming extinct. Changes to distribution of wildlife: migration patterns change which could affect local food chains and webs.

Carbon footprint and its reduction

1. The carbon footprint is the total amount of carbon dioxide and other greenhouse gases released into the atmosphere by human activity throughout the complete life cycle of a product, service or event. [1 mark]
2. **a** Solar panels convert energy from the Sun directly to electricity. [1 mark] Usually, producing electricity requires the burning of fossil fuels, which emits CO_2. If electricity is produced without burning fossil fuels, then CO_2 emissions will be reduced. [1 mark]

   b Solar panels only generate electricity during daylight [1 mark] and the amount they generate depends on the weather. [1 mark] The manufacturing of the actual solar panels has quite a large carbon footprint. [1 mark]
3. **a** 2007; [1 mark] this year has a large value from imported goods and services. [1 mark]

   b Imported goods and services. [1 mark] Acceptable levels of GHG emissions may be higher in different parts of the world/the UK imports much of the goods and services it uses/goods have further to travel. [1 mark]

   c It has stayed about the same between 1997 and 2013. [1 mark]

   d Any four points, but they must include the linked explanation.

   Double/triple glazed windows – reduces heat loss so less heating will be required. [1 mark] Loft /cavity wall installation – decreases heat loss, so less heating will be required. [1 mark] Switch appliances off when not using them – reducing the amount of electricity required. [1 mark] Install solar panels or other alternative energy sources – reduce the emission of greenhouse gases. [1 mark] Use low energy light bulbs – to reduce the amount of electricity required. [1 mark] Set the thermostat a bit lower – don't need to heat the house as much, so less heating will be required. [1 mark] Recycle products when finished with them – reducing GHG released when new products are made. [1 mark]

Air pollution from burning fuels

1.

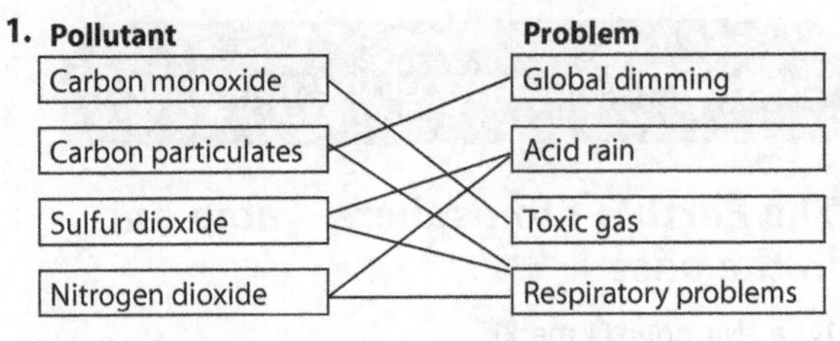

1 mark for the correct lines for each pollutant.

2.

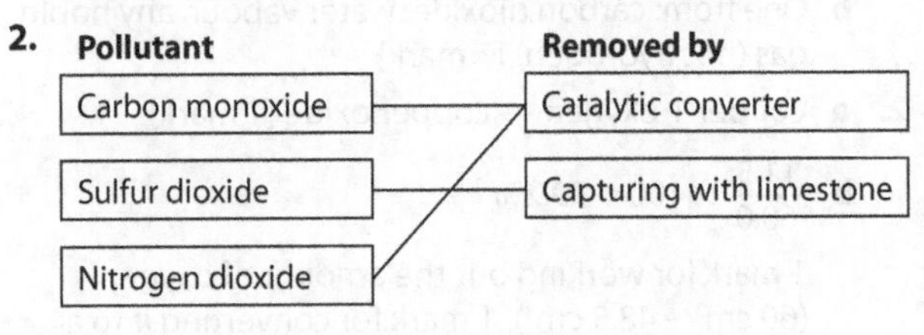

1 mark for the correct lines for each pollutant.

3. a High temperatures [1 mark]
   b $N_2(g) + 2O_2(g) \rightarrow 2NO_2(g)$
   1 mark for the correct formulae, 1 mark for the balancing, 1 mark for the state symbols.
   c Reaction A uses less oxygen than reaction B. [1 mark] Reaction B shows the complete combustion of heptane. All of the 7 carbon atoms present in the heptane molecule are oxidised to carbon dioxide molecules. [1 mark] Reaction A shows an example of the incomplete combustion of heptane. Only 1 carbon atom is fully oxidised to carbon dioxide, the rest form carbon monoxide. [1 mark] Therefore, carbon monoxide is formed when there is a limited oxygen supply so that incomplete combustion occurs. [1 mark]
   d Catalytic converters reduce the levels of nitrogen oxides and carbon monoxide released into the atmosphere from the car's exhaust. [1 mark]

Section 10: Using resources

What does the Earth provide for us?

1. Planting a tree to replace one that has been cut down; [1 mark] so that there are trees for future generations. [1 mark] (or any other suitable answer)
2. A renewable resource is one that can be replaced as quickly as it is used; [1 mark] whereas a finite resource is one that cannot be made again. [1 mark] Crude oil takes millions of years to form so once used up it cannot be replaced. Wood can be easily replaced as it comes from trees which can be grown in a few years. [1 mark]
3.

Marks awarded for this type of answer are determined by the Quality of Written Communication (QWC) as well as the standard of your scientific response.
0 marks: No relevant content.
Level 1 (1–2 marks): There is limited use of data in showing how changing diets may be linked to sustainable development.
Level 2 (3–4 marks) There is an explanation, supported by some data, showing how changing diets may be linked to sustainable development.
Level 3 (5–6 marks) There is a detailed explanation, supported by data, showing how changing diets may be linked to sustainable development.
Examples of the chemistry points made in the response: • Sustainable development is about meeting people's present needs without spoiling the environment for the future. • The water footprint for different foods is the number of litres of water used per litre or kilogram of foodstuff. • Foods such as meat have a high water footprint, e.g. beef 15 000, lamb 6000, pork 4500, chicken 3800 l of water per kg. • Foods such as cereals, fruit and vegetables have a much smaller water footprint than meat at around 1000 l of water per kg for barley and wheat; 250–500 l of water per kg for apples and oranges. • From 1960 to 2002 the amount of meat eaten per person in the USA has increased from about 85 to 122 kg per year. • Between 1980 and 2002 the amount of meat eaten per person in China has increased dramatically from about 15 to 55 kg per year. • The increase in meat consumption around the world since 1960 also means that has been an increase in water footprint in food production. • Water is vital to life and the supply of water must be conserved. If people continue to use too much water there may not be enough to go around. Also accept other arguments related to climate change such an increase in meat consumption leads to an increase in greenhouse gas emissions from cattle which may lead to global warming and climate change.

Safe drinking water

1. a Rain; also accept rivers, lakes or underground water [1 mark]
   b To allow solids to drop to the bottom. [1 mark] To remove fine particles from the water. [1 mark] To kill microbes. [1 mark]
2. a Heat the saltwater in a flask fitted with a condenser and a beaker to collect the pure water. [1 mark] As the water boils, it evaporates and passes into the condenser. [1 mark] Here the water vapour cools down and condenses back to liquid water. The water is then collected in a beaker. [1 mark]
   b It is too expensive because it requires a lot of energy to boil the water. [1 mark]
   c Pure water only contains water molecules and no other substances. [1 mark] Potable water is water that is safe to drink. [1 mark] It probably contains low levels of dissolved salts and small amounts of bacteria. [1 mark]
3. Measuring its boiling point. Pure water boils at exactly 100 °C at atmospheric pressure. [1 mark] As the concentration of dissolved salts or impurities in the water increases so does the boiling point. [1 mark]
   or Place some of the filtered water in an evaporating basin and heat until the water has all evaporated. [1 mark] Pure water will leave no solids in the basin, whereas impure water will. [1 mark]

Waste water treatment

1. The waste water may contain harmful bacteria that would make people ill/damage the environment. [1 mark] It may contain pollutants such as organic matter or harmful chemicals. [1 mark]

2. Anaerobic means without air/oxygen; aerobic means with air/oxygen. [1 mark]

Alternative methods of extracting metals

1. **a** Two from: High grade copper ore deposits have been used up. It is not economically viable to mine low grade copper ore. [1 mark] Mining has large environmental impact – noise, dust, destroys habitats. [1 mark] Metals such as copper are still needed. [1 mark]

   b phytomining, [1 mark] metal, [1 mark] harvested, [1 mark] burned [1 mark]

2. **a** Copper oxide [1 mark]

3. There are a lot of stages to the alternative methods which are both time and labour intensive, as plants have to be planted, grown and harvested. [1 mark]. There will be transport costs to the chemical plant where the extraction will take place [1 mark] and the cost of the actual extraction method. [1 mark]

4.

Marks awarded for this type of answer are determined by the Quality of Written Communication (QWC) as well as the standard of your scientific response.
0 marks: No relevant content.
Level 1 (1–2 marks): There is a brief description of methods used for metal extraction.
Level 2 (3–4 marks): There is explanation of why and how alternative methods are used.
Level 3 (5–6 marks): There is a detailed explanation of why and how alternative methods are used.
Examples of the chemistry points made in the response: • Earth's resources of metal ores are limited. • Many metal ores are of low grade. • Metal ores are becoming scarce and new ways of extracting the metals must be found to extract them. • Traditional mining methods of digging, moving and disposing of large amounts of rocks are no longer economically viable with low grade ore. • Bioleaching uses bacteria to produce leachate solutions that contain metal compounds. • Phytomining uses plants to absorb metal compounds from the soil. • Plants are harvested and then burnt to produce ash that contain the metal compounds – usually metal oxides. • In both bioleaching and phytomining, the metal compounds are processed by electrolysis or displacement reactions to produce the metal.

Life cycle assessment (LCA)

1. Extracting and processing raw materials, manufacture, use, disposal [1 mark]

2. **a** Accept either answer as correct as long as the answer given has the correct supporting arguments.

   The paper bag is best for the environment [1 mark] because the raw material is wood which is renewable. The raw materials for the plastic bag come from oil or gas which is a non-renewable source. [1 mark] Also less greenhouse gases are emitted during the life cycle of a paper bag, 0.23 kg of CO_2 equivalents compared to a plastic bag's 0.53 kg of CO_2 equivalents. [1 mark] Although slightly more energy is required to make the paper bag, 0.3 MJ, this energy could come from a renewable source and therefore have no further effect on the environment. [1 mark]

   Alternative answer

   The plastic bag is best for the environment [1 mark] because although the raw materials for the plastic bag come from oil or gas, which is a non-renewable source, when compared to the plastic bag it takes less energy to produce (0.3 MJ), [1 mark] produces a lot less solid waste (16 g as compared to 52 g) [1 mark] and the total emissions to the air are less than half (1.3 kg compared to 2.8 kg). [1 mark]

   b If the assessment is made by an employee of the manufacturing company it could be biased if the company wants to sell a lot of products. [1 mark]

   Alternative answer

   If the assessment is made by a pressure group it could be biased if the pressure group wants to stop a product being made. [1 mark]

3. The ceramic mugs have the lowest environmental impact, so they should be used for customers who want to buy coffee and drink it in the café. [1 mark] For customers on the go, travel mugs have a lower environmental impact than paper cups as long as the customers keep on re-using them. [1 mark] However, not everyone has a travel mug so I would suggest that the café sells travel mugs [1 mark] and charges slightly less for takeaway coffee when a travel mug is used instead of a paper cup. [1 mark]

   (Accept other versions as long as the points made are linked to the data provided.)

Ways of reducing the use of resources

1. **a** 35% [1 mark]

   b Two from: Glass, metal, paper some plastics or construction waste which could include bricks. [2 marks]

   c Separate waste into different materials to make it easier to recycle. [1 mark] Reduce the amount of waste by wasting less food/reusing plastic bags/bottles. [1 mark] or any other appropriate suggestions

2. 'Reduce' means to avoid waste, for example only buying what you actual need. [1 mark] 'Reuse' means to use an object again instead of throwing it away, for example a plastic bag. [1 mark] 'Recycle' is a method

for making new materials from ones that have already been used, for example recycling glass bottles so that new objects can be made from the glass. [1 mark]

3. a In 2000/01 only 11.2% of household waste was recycled compared to 43.7 % in 2016/17. [1 mark] There was a rapid increase in the rate of recycling between 2003/04 and 2010/11. [1 mark] Since 2011/12 the recycling rate has remained around 43.5%. [1 mark]

   b Two from: Recycling reduces the amount of waste going to landfill. [1 mark] Recycling reduces energy use during production of new materials from their raw materials. [1 mark] Recycling reduces the use of new raw materials/resources. [1 mark] It costs less to recycle and re-use the materials. [1 mark]

   c When materials are recycled, less raw materials are used; [1 mark] which reduces the pollution caused by extraction of materials from the Earth. For example, quarrying causes noise pollution and scars the environment. [1 mark] Often recycling uses less energy so less fossil fuels are being used to generate electricity, so reducing greenhouse gas emissions in to the air. [1 mark]. Less waste is being generated, which means less pollution caused at landfill sites. [1 mark]

Corrosion and its prevention

1. a The iron has reacted with oxygen (from the air) and water [1 mark] to form hydrated iron oxide, which is also known as rust. [1 mark]

   b The mass has decreased. [1 mark] Removing the rust means that some of the iron atoms are removed. [1 mark]

   c One from: paint them, cover them with a rubber sleeve, coat them with a layer of oil, store them in a dry place. [1 mark]

2. Aluminium reacts with oxygen from the air to form a layer of aluminium oxide on the surface of the can. [1 mark] This layer adheres/sticks to the aluminium underneath and protects it from further corrosion. [1 mark] In contrast, the layer of iron oxide (rust) separates from the layer of iron underneath, exposing further iron to corrosion. [1 mark] Aluminium cans cannot rust as rust is (hydrated) iron oxide and does not contain aluminium atoms.

3. a $2e^-$ [1 mark]

   b The oxygen gains electrons, which is the definition of reduction. [1 mark]

   c Magnesium is more reactive than zinc so it oxidises more readily than zinc. [1 mark]

   d The more reactive metal loses its electrons in preference to the iron. [1 mark]

   $Mg \rightarrow Mg^{2+} + 2e^-$ or $Zn \rightarrow Zn^{2+} + 2e^-$ [1 mark]

   Any Fe^{2+} ions are reduced back to Fe [1 mark]

   $Fe^{2+} + 2e^- \rightarrow Fe$ [1 mark]

   The iron is protected as the more reactive metal corrodes instead and can be easily replaced. [1 mark]

Alloys as useful materials

1. a An alloy is a metal made by mixing two or more metallic elements. [1 mark]

   b total number of atoms = 14 + 6 = 20

   $$\% \text{ of copper} = \frac{\text{number of copper atoms}}{\text{total number of atoms}} \times 100$$

   $$[1 \text{ mark}] = \frac{14}{20} \times 100 = 70\% \ [1 \text{ mark}]$$

2. a To make a helicopter, you want a material that is strong and light (has low density). [1 mark] Aluminium alloy has a strength of 600 MPa (which is the greatest of the materials given in the table) and a low density of 2.8 g/cm^3. [1 mark] Pure aluminium is slightly less dense (2.7 g/cm^3) but is not as strong (80 MPa). [1 mark] Iron and steel are quite strong but are too dense. [1 mark]

   b The metal will need to withstand high pressures. [1 mark] Steel has a tensile strength of 460 MPa whereas the tensile strength of iron is only 300 MPa; therefore steel is a better choice. [1 mark]

3. a Earrings made from 18 carat gold contain three times as much gold as those made from 6 carat gold. [1 mark] Gold is very expensive compared to the other metals it is alloyed with; [1 mark] so earrings with more gold will be more expensive.

Ceramics, polymers and composites

1. a Wood [1 mark]; it is made from long cellulose fibres held together by lignin. [1 mark] *or* concrete [1 mark]; it is made from cement with crushed aggregates or pebbles and sand. [1 mark]

   b Bricks are heat resistant/can withstand high temperatures/their melting point is over 1800 °C. [1 mark]

   c High density poly(ethene) has a higher melting point and higher density than low density poly(ethene). [1 mark] This is because in HDPE the polymer molecules line up regularly to give a crystalline structure but in LDPE side branches stop the polymer molecules lining up regularly. [1 mark] In HDPE the forces of attraction between the polymer molecules are stronger than in LDPE, meaning that more energy is required to overcome forces resulting in a higher melting point. [1 mark] In HDPE the crystalline structure means that there is a greater mass per unit volume than in LDPE, giving a higher density. [1 mark]

2.

Marks awarded for this type of answer are determined by the Quality of Written Communication (QWC) as well as the standard of your scientific response.
0 marks: No relevant content.
Level 1 (1–2 marks): There is a brief description of properties of thermosoftening and thermosetting polymers.
Level 2 (3–4 marks): There is a comparison of the structure and bonding of thermosoftening and thermosetting polymers.
Level 3 (5–6 marks): There is a detailed comparison of the structure and bonding of thermosoftening and thermosetting polymers.
Examples of the chemistry points made in the response: • Both types of polymers are very large molecules consisting of thousands of atoms. • Both polymers have repeating units and are made when the monomers join together. • In a thermosoftening polymer, the polymer chains can slide across each other making it flexible. • The forces of attraction between the polymer chains are weak and easily broken. • When heated the polymers can be reshaped as the polymer chains easily slide into different positions. • Thermosetting polymers are rigid, hard and brittle because the polymer chains are held in place by cross links. • Cross links form strong forces of attraction between the chains and a lot of energy is required to overcome these forces. • Thermosetting polymers do not soften when heated and cannot be re-shaped due to the strong forces between the chains.
(Accept also labelled diagrams.)

The Haber process

1. a $N_2(g) + 3H_2(g) \rightleftharpoons 2NH_3(g)$
   1 mark for each correct number, 1 mark for all molecules as gases
   b The reaction is reversible. [1 mark]
   c Increasing the pressure will favour the forward reaction (4 gas molecules form 2 gas molecules) [1 mark] and so more ammonia will be produced. [1 mark]
   d molecular mass of N_2 = 28, molecular mass of NH_3 = 17
   moles of $N_2 = \frac{375 \times 10^6}{28} = 13.39 \times 10^6$ [1 mark]
   mole ratio of N_2 to NH_3 is 1:2 as 1 mole of N_2 produces 2 moles of NH_3 [1 mark]
   moles of $NH_3 = 13.39 \times 10^6 \times 2 = 26.79 \times 10^6$
   mass of $NH_3 = 26.79 \times 10^6 \times 17 = 455.4 \times 10^6$ g or 455.4 tonnes
   mass of NH_3 = 455 tonnes to 3 sig fig. [1 mark]
2. a Accept 68–72% [1 mark]
   b From the graph, the yield decreases as the temperature increases. [1 mark] This implies that lower temperatures favour the forward reaction. [1 mark] According to Le Chatelier's Principle, the forward reaction must be exothermic (or the reverse reaction must be endothermic). [1 mark]
   c Only a very small percentage of the reactants are converted to ammonia at room temperature. [1 mark] The rate of reaction is also very slow. [1 mark]
   d Iron acts as a catalyst [1 mark] to increase the rate of reaction. [1 mark]

Production and uses of NPK fertilisers

1. The fertiliser contains 24% nitrogen, [1 mark] 6% phosphorus [1 mark] and 6% potassium. [1 mark]
2. Different fertilisers have different N:P:K values, as some contain just one of these important elements and some contain two. To provide all three elements requires the use of at least two different fertilisers. [1 mark] Different crops take different proportions of different elements out of the soil which need to be replaced by fertilisers. [1 mark] For example, if the soil is deficient in phosphorus then the farmer would use ammonium phosphate; [1 mark] but if it was deficient in potassium the framer would have to use potassium sulfate. [1 mark] (Accept other examples as long as they are correct.)
3. hydrochloric acid + potassium hydroxide → potassium chloride + water [1 mark] Measure out the potassium hydroxide solution with a measuring cylinder and pour it into a conical flask. [1 mark] Using a burette, slowly add hydrochloric acid to the flask until the mixture is neutral. You will need to use a pH probe to show this. [1 mark] Pour the solution into an evaporating basin and heat gently until crystals form. Filter off the crystals using a filter funnel and paper. Leave the crystals to dry. [1 mark]
4. a ammonia + nitric acid → ammonium nitrate + water
   1 mark for correct reactants, 1 mark for correct products.
   b Hydrogen, made from methane, is reacting with nitrogen from the air [1 mark] to produce ammonia. [1 mark]